Recent Advances in Space Debris

Recent Advances in Space Debris

Editors

**Lorenzo Olivieri
Kanjuro Makihara
Leonardo Barilaro**

Basel • Beijing • Wuhan • Barcelona • Belgrade • Novi Sad • Cluj • Manchester

Editors

Lorenzo Olivieri
CISAS "Giuseppe Colombo"
Università degli Studi di
Padova
Padova
Italy

Kanjuro Makihara
Department of Aerospace
Engineering
Tohoku University
Sendai
Japan

Leonardo Barilaro
Institute of Engineering and
Transport
The Malta College of Arts,
Science & Technology
Paola
Malta

Editorial Office
MDPI
St. Alban-Anlage 66
4052 Basel, Switzerland

This is a reprint of articles from the Special Issue published online in the open access journal *Applied Sciences* (ISSN 2076-3417) (available at: https://www.mdpi.com/journal/applsci/special_issues/Recent_Advances_in_Space_Debris).

For citation purposes, cite each article independently as indicated on the article page online and as indicated below:

Lastname, A.A.; Lastname, B.B. Article Title. *Journal Name* **Year**, *Volume Number*, Page Range.

ISBN 978-3-7258-0436-8 (Hbk)
ISBN 978-3-7258-0435-1 (PDF)
doi.org/10.3390/books978-3-7258-0435-1

© 2024 by the authors. Articles in this book are Open Access and distributed under the Creative Commons Attribution (CC BY) license. The book as a whole is distributed by MDPI under the terms and conditions of the Creative Commons Attribution-NonCommercial-NoDerivs (CC BY-NC-ND) license.

Contents

About the Editors vii

Preface ix

Lorenzo Olivieri, Kanjuro Makihara and Leonardo Barilaro
Editorial for Special Issue: Recent Advances in Space Debris
Reprinted from: *Appl. Sci.* **2024**, *14*, 954, doi:10.3390/app14030954 1

Lorenzo Olivieri, Cinzia Giacomuzzo, Stefano Lopresti and Alessandro Francesconi
Research at the University of Padova in the Field of Space Debris Impacts against Satellites: An Overview of Activities in the Last 10 Years
Reprinted from: *Appl. Sci.* **2023**, *13*, 3874, doi:10.3390/app13063874 4

Leonardo Barilaro, Mark Wylie and Theeba Shafeeg
Design of the Sabot-Stopping System for a Single-Stage Light-Gas Gun for High-Velocity Impacts
Reprinted from: *Appl. Sci.* **2023**, *13*, 7664, doi:10.3390/app13137664 20

Alexander Kraus, Andrey Buzyurkin, Ivan Shabalin and Evgeny Kraus
Numerical Modelling of High-Speed Loading of Periodic Interpenetrating Heterogeneous Media with Adapted Mesostructure
Reprinted from: *Appl. Sci.* **2023**, *13*, 7187, doi:10.3390/app13127187 34

William P. Schonberg
Extending the NNO Ballistic Limit Equation to Foam-Filled Dual-Wall Systems
Reprinted from: *Appl. Sci.* **2023**, *13*, 800, doi:10.3390/app13020800 56

Xiangxu Lei, Shengfu Xia, Hongkang Liu, Xiaozhen Wang, Zhenwei Li, Baomin Han, et al.
An Improved Range-Searching Initial Orbit-Determination Method and Correlation of Optical Observations for Space Debris
Reprinted from: *Appl. Sci.* **2023**, *13*, 13224, doi:10.3390/app132413224 67

Gongqiang Li, Jing Liu, Hai Jiang and Chengzhi Liu
Research on the Efficient Space Debris Observation Method Based on Optical Satellite Constellations
Reprinted from: *Appl. Sci.* **2023**, *13*, 4127, doi:10.3390/app13074127 81

Liang Hu, Dianqi Sun, Huixian Duan, An Shu, Shanshan Zhou and Haodong Pei
Non-Cooperative Spacecraft Pose Measurement with Binocular Camera and TOF Camera Collaboration
Reprinted from: *Appl. Sci.* **2023**, *13*, 1420, doi:10.3390/app13031420 98

Maxime Hubert Delisle, Olga-Orsalia Christidi-Loumpasefski, Barış C. Yalçın, Xiao Li, Miguel Olivares-Mendez and Carol Martinez
Hybrid-Compliant System for Soft Capture of Uncooperative Space Debris
Reprinted from: *Appl. Sci.* **2023**, *13*, 7968, doi:10.3390/app13137968 119

Francesco Barato
Comparison between Different Re-Entry Technologies for Debris Mitigation in LEO
Reprinted from: *Appl. Sci.* **2022**, *12*, 9961, doi:10.3390/app12199961 145

Kohei Takeda, Toshinori Kuwahara, Takumi Saito, Shinya Fujita, Yoshihiko Shibuya, Hiromune Ishii, et al.
De-Orbit Maneuver Demonstration Results of Micro-Satellite ALE-1 with a Separable Drag Sail
Reprinted from: *Appl. Sci.* **2023**, *13*, 7737, doi:10.3390/app13137737 . . . . **185**

About the Editors

Lorenzo Olivieri

Lorenzo OLIVIERI is a postdoctoral researcher at the Centre for Space Studies CISAS G. Colombo at the University of Padova. He graduated in Aerospace Engineering in 2011 and received a Ph.D. in Measures for Space from the University of Padova in 2015. He is adjunct lecturer at the Department of Industrial Engineering of the University of Padova.

His research interests include small satellite technologies, space debris protection, as well as drones and balloon systems. Throughout his career, he worked in international teams as part of the ReDSHIFT and E.T.PACK projects in the framework of the European H2020 program. Currently, he is a member of the Italian Space Agency delegation for the Working Group 3 "Protection" at the Inter-Agency Debris Coordination Committee (IADC).

Kanjuro Makihara

Kanjuro MAKIHARA, Doctor of Engineering, received his Bachelor's degree in Aeronautics and Astronautics from the University of Tokyo in 1998 and completed his Ph.D. program from the Graduate School of the University of Tokyo in 2004. Since 2004, he was an Aerospace Project Research Associate at JAXA/ISAS and has devoted himself to energy-recycling vibration suppression for space structures. After serving as a visiting researcher at the University of Cambridge, U.K., he has worked as an Associate Professor of Aerospace Engineering at the Tohoku University since 2011, and, in 2019, he became a professor at the Tohoku University. His current research interests involve semi-active vibration suppression, self-powered energy-harvesting, dynamics of flexible structures, and issues pertaining to space debris.

Leonardo Barilaro

Dr Leonardo BARILARO, also known as the "Space Pianist", is a Senior Lecturer in Aerospace Engineering at MCAST (Malta). With a PhD in Sciences, Technologies, and Measurements for Space from the University of Padova (Italy), he has conducted extensive research on new techniques to assess and mitigate space debris risks. Currently, he is pursuing new research projects in the areas of hypervelocity impacts and aerospace structures. Dr Barilaro's technical background encompasses mechanical measurements for engineering, constructions, structures, fluid dynamics, and aerospace systems, with a strong background in finite element analysis (FEA) simulations.

As a polymath artist, he merges his passion for music and science to promote space exploration and mitigate the impact of the climate crisis. He has produced multiple studio albums and released a daily space music track throughout 2022. His composition "Maleth" was featured onboard the International Space Station (ISS) during the SpaceX CSR25 mission and was broadcast to Earth in August 2022. In 2023, he further expanded his reach with another composition onboard the ISS with the Maleth 3 project.

Dr Barilaro's fusion of music and science aims to inspire audiences, transcend cultural barriers, and connect people to the wonders of Space.

Preface

Space debris has been identified as an actual hazard for operational satellites, for human activities in space, and generally for the exploitation of near-Earth orbits. They can be generated by various sources, including in orbit operations, the breakup of defunct satellites and rocket stages, the degradation of orbital elements, and collision events.

The potential for collisions poses risks to autonomous satellites, crewed spacecraft, and the International Space Station. Moreover, the long-term sustainability of space activities is jeopardized as the number of debris continues to rise. As a result, space agencies, scientists, and policymakers are actively engaged in efforts to monitor, mitigate, and find solutions to address the challenges posed by space debris.

In this context, understanding the causes, consequences, and potential solutions to the issue of space debris is crucial for maintaining a sustainable near-Earth orbital environment. This Special Issue delves into the topic by addressing three of the main investigative lines: (1) understanding the physical processes behind in-orbit fragmentation; (2) assessing the in-orbit population of space debris; and (3) developing mitigation strategies and enabling technologies that are used to remove end-of-life satellites and large relicts from orbit.

Lorenzo Olivieri, Kanjuro Makihara, and Leonardo Barilaro
Editors

Editorial

Editorial for Special Issue: Recent Advances in Space Debris

Lorenzo Olivieri [1,*], Kanjuro Makihara [2] and Leonardo Barilaro [3]

1. CISAS "G. Colombo", University of Padova, Via Venezia 15, 35131 Padova, PD, Italy
2. Department of Aerospace Engineering, Tohoku University, Sendai 980-8579, Japan; kanjuro.makihara.e3@tohoku.ac.jp
3. Department of Aviation, The Malta College of Arts, Science & Technology, Triq Kordin, PLA 9032 Paola, Malta; leonardo.barilaro@mcast.edu.mt
* Correspondence: lorenzo.olivieri@unipd.it

The near-Earth space debris environment represents an existing hazard for human activities in space. The increasing number of man-made objects resident in orbit leads to a growing risk of collisions involving active spacecraft, which could cause anything from the loss of important functionalities to vehicle break-up and, in parallel, the fragmentation of satellites that are no longer operational. The scientific community worries that such a process may lead to large fragmentation events and a cascade effect that would prevent the safe access and exploitation of entire orbital regions.

Addressing the space debris problem and finding potential mitigation solutions is a challenge that requires a holistic approach and the collaboration of all involved stakeholders. It is of paramount importance to clarify the mechanisms that lead to the generation of space debris and their distribution at different altitudes, especially in crowded orbits, and to find strategies by which to remove potential sources of novel debris (e.g., end-of-life satellites, spent rocket stages). In this Special Issue, three of the main investigative lines concerning space debris are presented: (1) understanding the physical processes behind in-orbit fragmentation; (2) assessing the in-orbit population of space debris; and (3) developing mitigation strategies and enabling technologies by which to remove end-of-life satellites and large relicts from non-operational spacecraft.

The direct observation of space debris collisions in space is extremely difficult; for this reason, on-ground impact testing and numerical simulations are the most frequently employed methods by which to investigate the relevant fragmentation physics and to evaluate the survivability of space shields and structures. In this context, in the Special Issue's feature paper [1], the authors present a review of the experimental and simulation activities performed in a research laboratory, describing the main findings and underlining the importance of such activities for better understanding the space debris problem. The importance of experimental facilities is also addressed in [2], where the authors introduce advances in the technologies currently employed in hypervelocity testing. Both the complexity and the advantages of numerical simulations are well addressed in [3]; the authors describe the analysis of heterogeneous materials subjected to impacts and present the simulation of single- and multiple-space-debris impacts. To quantitatively assess the survivability of spacecraft structures after impact, ballistic limit equations are often employed; these represent a fundamental tool in the risk assessment and design of spacecraft protection. In [4], the authors present the extension of BLEs for foam-filled dual-wall systems, showing how the comparison between test data and numerical simulations can lead to a marked improvement in the prediction capability of such a useful tool.

Understanding the physics of space debris generation can help in defining and modeling the future trends of the space debris environment; however, the investigation and cataloguing of the current population is a fundamental step in assessing risks and suggest mitigation strategies. Observations can be performed both from large ground facilities and via distributed in-orbit systems. In [5], the authors provide an improved method for

Citation: Olivieri, L.; Makihara, K.; Barilaro, L. Editorial for Special Issue: Recent Advances in Space Debris. *Appl. Sci.* **2024**, *14*, 954. https://doi.org/10.3390/app14030954

Received: 18 December 2023
Accepted: 19 January 2024
Published: 23 January 2024

Copyright: © 2024 by the authors. Licensee MDPI, Basel, Switzerland. This article is an open access article distributed under the terms and conditions of the Creative Commons Attribution (CC BY) license (https://creativecommons.org/licenses/by/4.0/).

determining the orbital parameters of space objects obtained via very short arc observations from a ground telescope; it is shown that this method has a high success rate and can be employed for the rapid assessment of fragmentation events. With respect to in situ observation, in [6], a model for designing and assessing the efficiency of a constellation for space debris observation is presented; through long-term continuous observation, these constellations could maintain an orbit catalogue of the majority of the objects in LEO, and they could also provide up-to-date information for space situational awareness.

In addition to collision modeling and population cataloguing, active ons shall be taken in order to mitigate the risk of further pollution in Earth's orbit. Removing spacecraft at the end of their operational life, or when otherwise malfunctioning, requires state-of-the-art technologies and complex mission architectures. First, cooperative or non-cooperative targets shall be safely approached and observed in order that we might assess their structural integrity, status, and attitude. In [7], an architecture based on binocular and time-of-flight cameras is implemented to reconstruct the pose of an uncontrolled target, and experimental results indicate good accuracy in reconstructing the pose, with position errors within 1 cm and angular errors below 1 deg for low-speed tumbling. Another complex task is the creation of a mechanical joint between the target and a servicing or deorbiting module. In [8], a versatile interface that can also be fit in CubeSat-sized vehicles is presented, and it effectiveness in performing soft-capturing with uncooperative targets is assessed. The removal of space objects requires the performance of orbital maneuvers, which can be performed by active and passive systems. In [9], a comparison of existing strategies is performed, indicating the strength and the limit of the different approaches. In addition, the author suggests that for LEO satellites, low-thrust propulsion combined with drag augmentation systems could be an effective and low-cost solution for both drag compensation during operations and controlled re-entry at end of life. Among drag augmentation devices, drag sails already represent the state of the art. In [10], the authors present an in-orbit demonstration performed by a micro-satellite equipped with a sail that, despite a few subsystem failures, was still capable of lowering the spacecraft altitude from 500 km to 400 km.

The collection of papers in this Special Issue represent the state of the art in space debris research. Addressing this issue with competent and effective strategies is a complex challenge for all of the involved stakeholders; as editors of this Special Issue, we hope that the works published herein will increase public awareness of, and stimulate further research on, this captivating and crucial topic.

Acknowledgments: We wish to thank the authors contributing to this Special Issue. Their effort in presenting state-of-the-art works was paramount to the success of this Special Issue, and the high quality of the submitted works underlines the importance of researching the physical processes of space debris generation and the mitigation strategies necessary to maintaining safe access to Earth's orbits.

Conflicts of Interest: The authors declare no conflicts of interest.

References

1. Olivieri, L.; Giacomuzzo, C.; Lopresti, S.; Francesconi, A. Research at the University of Padova in the Field of Space Debris Impacts against Satellites: An Overview of Activities in the Last 10 Years. *Appl. Sci.* **2023**, *13*, 3874. [CrossRef]
2. Barilaro, L.; Wylie, M.; Shafeeg, T. Design of the Sabot-Stopping System for a Single-Stage Light-Gas Gun for High-Velocity Impacts. *Appl. Sci.* **2023**, *13*, 7664. [CrossRef]
3. Kraus, A.; Buzyurkin, A.; Shabalin, I.; Kraus, E. Numerical Modelling of High-Speed Loading of Periodic Interpenetrating Heterogeneous Media with Adapted Mesostructure. *Appl. Sci.* **2023**, *13*, 7187. [CrossRef]
4. Schonberg, W.P. Extending the NNO Ballistic Limit Equation to Foam-Filled Dual-Wall Systems. *Appl. Sci.* **2023**, *13*, 800. [CrossRef]
5. Lei, X.; Xia, S.; Liu, H.; Wang, X.; Li, Z.; Han, B.; Sang, J.; Zhao, Y.; Luo, H. An Improved Range-Searching Initial Orbit-Determination Method and Correlation of Optical Observations for Space Debris. *Appl. Sci.* **2023**, *13*, 13224. [CrossRef]
6. Li, G.; Liu, J.; Jiang, H.; Liu, C. Research on the Efficient Space Debris Observation Method Based on Optical Satellite Constellations. *Appl. Sci.* **2023**, *13*, 4127. [CrossRef]

7. Hu, L.; Sun, D.; Duan, H.; Shu, A.; Zhou, S.; Pei, H. Non-Cooperative Spacecraft Pose Measurement with Binocular Camera and TOF Camera Collaboration. *Appl. Sci.* **2023**, *13*, 1420. [CrossRef]
8. Hubert Delisle, M.; Christidi-Loumpasefski, O.O.; Yalçın, B.C.; Li, X.; Olivares-Mendez, M.; Martinez, C. Hybrid-Compliant System for Soft Capture of Uncooperative Space Debris. *Appl. Sci.* **2023**, *13*, 7968. [CrossRef]
9. Barato, F. Comparison between Different Re-Entry Technologies for Debris Mitigation in LEO. *Appl. Sci.* **2022**, *12*, 9961. [CrossRef]
10. Takeda, K.; Kuwahara, T.; Saito, T.; Fujita, S.; Shibuya, Y.; Ishii, H.; Okajima, L.; Kaneko, T. De-Orbit Maneuver Demonstration Results of Micro-Satellite ALE-1 with a Separable Drag Sail. *Appl. Sci.* **2023**, *13*, 7737. [CrossRef]

Disclaimer/Publisher's Note: The statements, opinions and data contained in all publications are solely those of the individual author(s) and contributor(s) and not of MDPI and/or the editor(s). MDPI and/or the editor(s) disclaim responsibility for any injury to people or property resulting from any ideas, methods, instructions or products referred to in the content.

Review

Research at the University of Padova in the Field of Space Debris Impacts against Satellites: An Overview of Activities in the Last 10 Years

Lorenzo Olivieri [1,*], Cinzia Giacomuzzo [1], Stefano Lopresti [1] and Alessandro Francesconi [2]

1 CISAS "G. Colombo", University of Padova, Via Venezia 15, 35131 Padova, PD, Italy
2 CISAS "G. Colombo"-DII, University of Padova, Via Venezia 1, 35131 Padova, PD, Italy
* Correspondence: lorenzo.olivieri@unipd.it; Tel.: +39-049-8276837

Abstract: Space debris represent a threat to satellites in orbit around Earth. In the case of impact, satellites can be subjected to damage spanning from localized craterization to subsystem failure, to complete loss of the vehicle; large collision events may lead to fragmentation of the spacecraft. Simulating and testing debris impacts may help in understanding the physics behind these events, modelling the effects, and developing dedicated protection systems and mitigation strategies. In this context, the Space Debris group at the University of Padova investigates in-space collisions with experimental campaigns performed in a dedicated Hypervelocity Impact Facility and with numerical simulations with commercial and custom software. In this paper, an overview is given of the last 10 years of research activities performed at the University of Padova. First, the hypervelocity impact testing facility is described and the main experimental campaigns performed in the last few years are summarized. The second part of this work describes impact modelling research advances, focusing on the simulation of complex collision scenarios.

Keywords: space debris; hypervelocity impact; spacecraft fragmentation; impact modelling

Citation: Olivieri, L.; Giacomuzzo, C.; Lopresti, S.; Francesconi, A. Research at the University of Padova in the Field of Space Debris Impacts against Satellites: An Overview of Activities in the Last 10 Years. *Appl. Sci.* **2023**, *13*, 3874. https://doi.org/10.3390/app13063874

Academic Editor: Jérôme Morio

Received: 1 March 2023
Revised: 15 March 2023
Accepted: 16 March 2023
Published: 18 March 2023

Copyright: © 2023 by the authors. Licensee MDPI, Basel, Switzerland. This article is an open access article distributed under the terms and conditions of the Creative Commons Attribution (CC BY) license (https:// creativecommons.org/licenses/by/ 4.0/).

1. Introduction

In recent years, the number of human-made objects launched into space has continued to grow [1–3], increasing the danger of the near-Earth debris environment and increasing the risk of in-space collisions [4–7]. Mitigation regulations [8] and strategies [9] have been and are continuously being proposed, and active [10] and passive [11,12] debris removal and post-mission disposal technologies are under development. It is clear that the risk of debris impacts shall be considered in the development of a space mission: a collision can cause damage to spacecraft subsystems and affect their functionality [13], up to the complete loss of the vehicle. Eventually, critical events may lead to the partial or complete fragmentation of the impacted bodies [14].

In this context, on one hand, dedicated protections are designed to shield satellites from hypervelocity impacts; ground tests are usually employed to assess their capability to mitigate collisions [15,16] and to protect spacecraft components [17,18]. Furthermore, critical elements are usually subjected to similar tests to evaluate their survivability in the debris environment [19]. On the other hand, it is crucial to understand the physical processes involved in spacecraft collisions through numerical simulations and ground-based experiments. Currently, spacecraft collisions and large fragmentation events are evaluated mostly with empirical or semi-empirical tools (e.g., NASA SBM [20], FASTT [21,22], IMPACT [23,24]), as numerical simulations based on the impact physics (e.g., hydrocodes) are usually too complex and resource-consuming to investigate a wide range of collision scenarios. Ground tests of complex satellite models [25–27] usually require large impact facilities and a large amount of worker hours for fragment collection and analysis [28].

The Space Debris group of the University of Padova has investigated both these topics in collaboration with national and international partners; this paper presents the experimental and numerical research activities performed by the group in the last 10 years. The next section introduces the Hypervelocity Impact Facility employed for the experimental investigation, consisting of a two-stage Light-Gas Gun capable of accelerating projectiles up to 100 mg at a maximum speed of 5.5 km/s. The purposes of the experimental campaigns, which are presented in Section 3, are (i) the evaluation of the ballistic response of satellite structures and components subjected to impacts with small debris of mm size, (ii) the development of new shields with enhanced shielding capabilities, (iii) the investigation of fragmentation processes at component and system level, and (iv) the development of smart systems for in situ small debris impact detection. The second part of this work describes the impact modelling research (Section 4), focusing on simulations of collision scenarios involving both basic elements and full satellites (Section 5). In particular, the collected results can be compared to those of current breakup models such as the NASA SBM, and to derive new semi-empirical formulations of the fragmentation process.

2. The Hypervelocity Impact Facility

The Center for Space Activities CISAS "G. Colombo" was founded in 1991 as an interdisciplinary structure able to host research activities performed by the different departments of the University of Padova in the field of space science and technology. Regarding space debris, the activities grew considerably in the late 1990s with the development of a unique impact facility based upon a Two-Stage Light-Gas Gun (LGG). From 2001, a complete laboratory was established for high-velocity and hypervelocity impact physics, with the development and procurement of various instruments for impact diagnostics [29–32]. The current facility can be seen in Figure 1 (left); it is capable of accelerating projectiles up to 100 mg at a maximum speed of 5.5 km/s; its main innovation is the high shot frequency, up to 10 experiments per day, achievable due to a specific setup which employs reusable components in the whole main gas gun subsystem, including the piston and the high-pressure section. The current LGG consists of (1) a first-stage reservoir, where a light gas is stored at high pressure, (2) a pneumatic piston, which is accelerated by the gas in the first stage, (3) a tube acting as a second stage in which a light gas is compressed by the piston, (4) a high-pressure section with reusable pneumatic valves to release the high-pressure gas on the back of the launch package, (5) a barrel, in which the launch package, made by the sabot and the projectile, is accelerated by the compressed gas from the second stage, (6) a flight chamber in which the projectile continues its trajectory towards the target, (7) a sabot stopping system, and (8) an impact chamber in which the projectile hits the target. The projectile velocity is measured by two laser optical barriers placed just before the impact chamber (Figure 1, right); a high-velocity camera is placed on one side of the impact chamber and is employed to record impact videos. In addition, dedicated sensors can be added to instrument the target depending on the experimental setup (e.g., accelerometers, acoustic sensors, ballistic pendulum).

 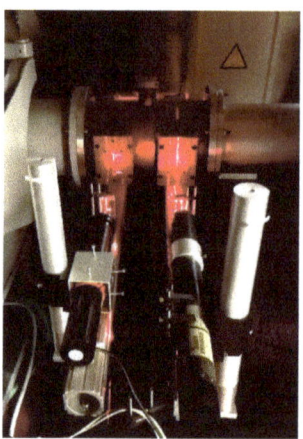

Figure 1. CISAS Hypervelocity Impact Facility with the LGG (**left**) and laser system for projectile velocity measurement (**right**).

3. Experimental Activities

In this section, the most important experimental activities performed in the last decade with the Hypervelocity Impact Facility of the University of Padova are summarized. They are divided into four main branches: (i) evaluation of the ballistic response of satellite systems subjected to impacts with small debris of mm size, (ii) development of new shields with enhanced shielding capabilities, (iii) investigation of the fragmentation process at component and system level, and (iv) development of smart systems for in situ small debris impact detection.

3.1. Evaluation of the Ballistic Response of Satellite Systems

The ballistic response of structures and systems is usually investigated through hypervelocity impact tests and often through simulations to explore impact ranges not achievable in the laboratory. These studies aim to identify the projectile critical diameter (usually, in respect of the impact velocity and/or other impact parameters) which makes a structure or a system fail in the case of a collision. The ballistic response can be expressed through Ballistic Limit Equations (BLEs). In this context, the investigations performed by the Space Debris group of the University of Padova focused on the BLEs for tape tethers.

Tape Tethers

Space tethers have been proposed for a large number of applications, from futuristic space elevators to dynamic systems able to raise or lower the orbit of a satellite [33]. In particular, electrodynamic tethers employ the interaction with the space plasma and the Earth's magnetosphere to generate thrusts that can be used for orbital maneuvers. Among the main constraints in this application, the scientific community is worried that long and thin wires could be strongly affected by the debris environment: any small impact may result in the severing of the tether [34]. To overcome this limitation, thin tape tethers were proposed due to their non-symmetric behavior during impacts (in addition to other positive features such as an enhanced capacity to collect plasma for electrodynamic applications). The investigation of the ballistic response of tape tethers to impacts was carried out at the University of Padova in the framework of Project 262972 (BETs) funded by the European Commission under the FP7 Space Program [35,36]. In this activity, a set of 24 experiments with different projectiles diameters, velocities, and impact angles was performed on aluminum and polymeric tapes; 112 additional numerical simulations were executed with a commercial software ANSYS™ Autodyn to evaluate impact conditions outside of the LGG operational range. The experimental setup and the damage on a PEEK tape from an impact test can be seen in Figure 2; Figure 3, which report the ballistic limit of the tape tether in respect of the impact angle and velocity as obtained from the analysis of the test results. It is shown that there is a minimum value of debris velocity v* below which no critical damage is possible; furthermore, there is a minimum velocity-dependent value d* of debris diameter below which no critical damage is possible. This is an extremely important result, since it sets a minimum particle diameter for risk assessment and thus excludes a large part of the flux from risk computations [35].

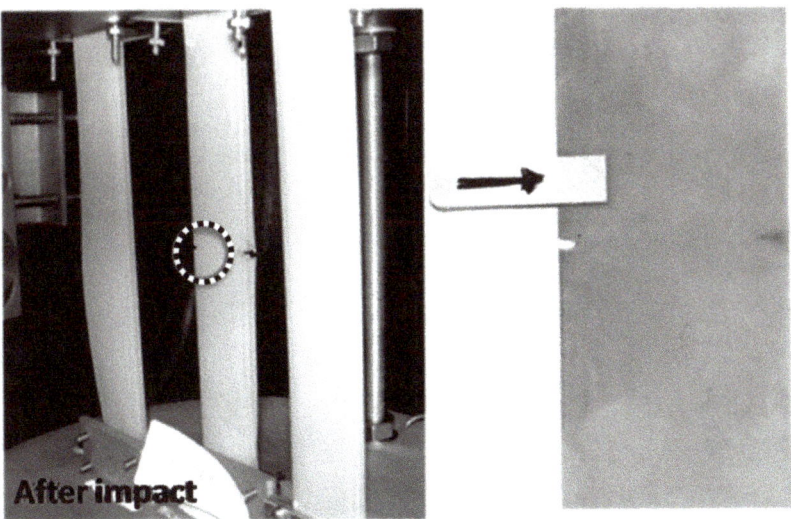

Figure 2. Setup for tape tether impact tests (**left**) and impact damage due to a tape tether (**right**, impact direction parallel to the tether face).

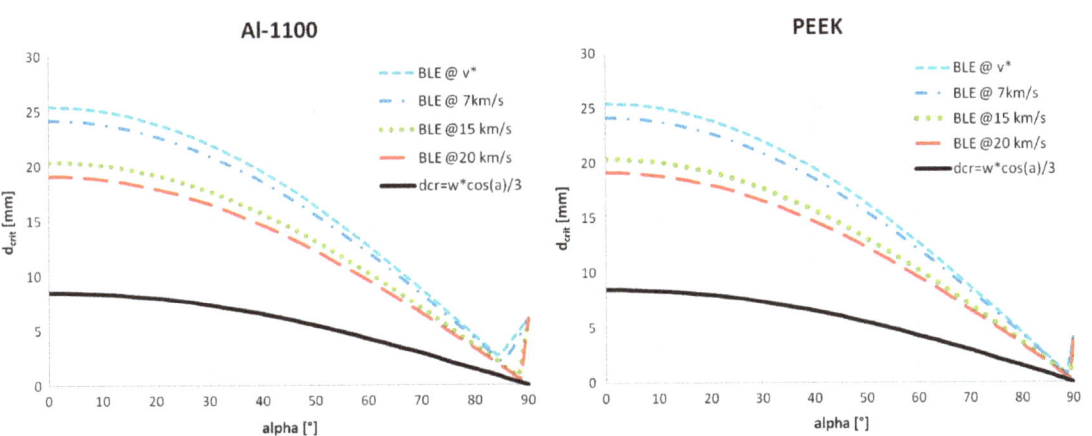

Figure 3. Aluminum (**left**) and polymeric (PEEK, **right**) tape tethers' ballistic limit in respect of the impact angle (alpha) and impact velocity; d_{CRIT} represents the critical impacting particle diameter.

The advantage of employing tape tethers has become clear to the scientific community. The European Commission has funded the current projects E.T.Pack and E.T. PACK-Fly, which aim to demonstrate tether technology for satellites deorbiting at the end of their operational life [12,37]; both projects employ the same tape tethers investigated in the BETs study.

3.2. Advanced Shielding for Spacecraft

Among the experimental activities, test campaigns were dedicated to the investigation of the shielding capability of advanced configurations or novel materials. With respect to standard sandwich panels, currently used as structural elements with a certain shielding capability, new materials and configurations may show enhanced performances. Hypervelocity impact testing allows the determination of the ballistic limit of these new configurations, as well as the ability to fragment impacting bodies to further reduce the vulnerability of internal components to perforation.

3.2.1. Carbon-Fiber-Reinforced Panels (CFRPs)

In the framework of a collaboration between CISAS and JAXA (Japan Aerospace Exploration Agency), a campaign of hypervelocity tests was performed on CFRPs (Figure 4). In total, 45 experiments were performed on panels 2.3, 3.5, and 4.3 mm thick, with spherical projectiles with diameter from 0.8 to 2.9 mm impacting at a speed of 2–5 km/s. This campaign allowed the definition of the panels' Ballistic Limit Equation, as well as the determination of the panels' internal delamination during hypervelocity impacts. Further data can be found in [38].

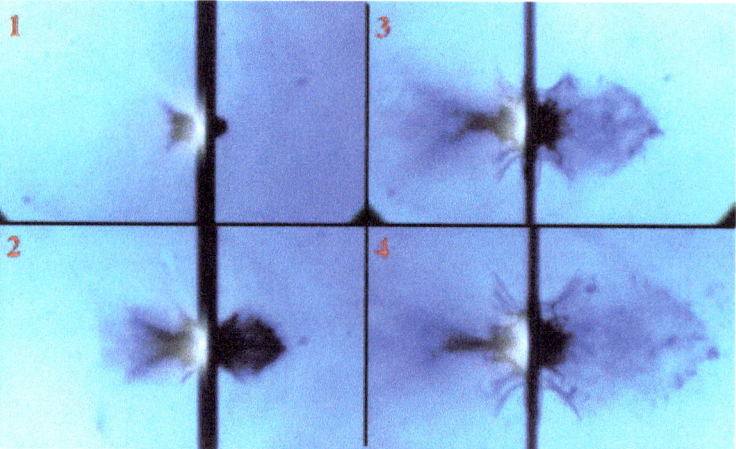

Figure 4. Frames from the video of an impact on CFRP panels. The projectile is impacting from the left; two debris clouds are generated respectively from the panel front and back faces.

3.2.2. Self-Healing Advanced Panels

In the framework of the IMpact BEhavior of MUltifunctional materialS (IMBEMUS) project, supported by the CARIPARO Foundation, the mechanical behavior of multifunctional panels was studied under impact conditions. The investigation required both high-velocity impact experiments and numerical simulations, performed with ANSYS™ Autodyn. In particular, multi-layer panels composed of ionomeric polymers with self-healing properties were subjected to 14 impact experiments; the self-healing (see Figure 5) was successful in all but one ionomer samples and the primary damage on ionomeric polymers was found to be significantly lower than that on aluminum. On the other hand, aluminum plates exhibited slightly better debris fragmentation abilities, even though the protecting performance of ionomers seemed to improve at increasing impact speed. Further data and the main experimental campaign results can be found in [39–41].

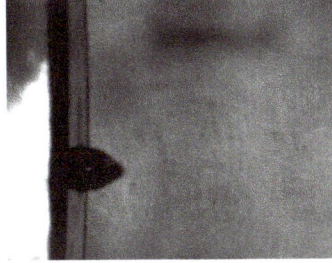 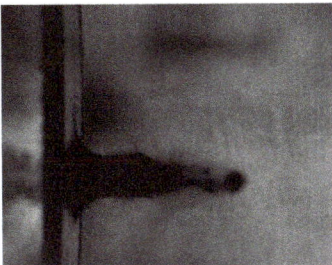 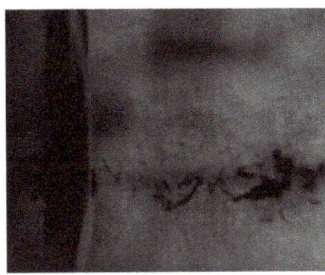

Figure 5. Frames from impact test video showing the panel's self-healing capability.

3.2.3. 3D-Printed Shields

The H2020 ReDSHIFT (Revolutionary Design of Spacecraft through Holistic Integration of Future Technologies) project focused on the development of highly innovative and low-cost spacecraft solutions for debris protection and mitigation, including design for demise and novel manufacturing strategies and technologies [42]. In this project, hypervelocity impact testing investigated the response of aluminum shield panels with complex core geometry produced through additive manufacturing. Both simple plates (Figure 6, left) and multi-layer configurations (Figure 6, right) were investigated in the experimental campaign, which included 20 shots with different projectiles at velocities in the range between 1 and 5 km/s. Results showed that the manufacturing process did not significantly influence the response of samples to impacts and that multi-layer configurations present improvements in terms of debris shielding with respect to standard honeycomb sandwich panels with equivalent areal density and thickness [43].

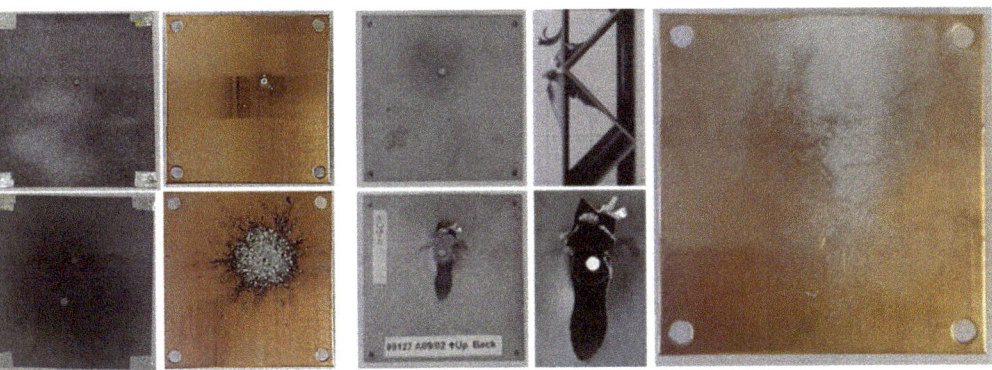

Figure 6. Pictures of 3D-printed samples subjected to impact tests and the copper witness plates placed behind them [43]. On the left, impact tests on simple plates (2.9 mm projectile at 1 km/s, top, and 5 km/s, bottom). On the right, test on a multi-layer corrugated panel (2.9 mm at 4.8 km/s).

3.3. Investigation of the Fragmentation Process

Investigating the physics behind orbital collisions, and in particular the production of new fragments, is important for understanding the evolution of the space debris environment. In this context, the University of Padova is studying the fragmentation process of both simple and complex targets, with particular focus on the characteristics and the distributions of the generated debris. This work, supported by the European Space Agency in the framework of a collaboration with SpaceDys [44], included a campaign of impact tests on aluminum and CFRP simple plates and sandwich panels and a catastrophic fragmentation experiment on a Picosatellite mock-up.

3.3.1. Aluminum and CFRP Plates

The investigation of aluminum thin plates aimed to evaluate the effect of impact velocity and target thickness on fragment generation. Four tests were performed on aluminum plates of two different thicknesses (2 and 5 mm). Aluminum spherical projectiles of 2.9 mm in diameter were launched at velocities between 3.5 and 4.8 km/s. The fragments generated by the impact were collected, catalogued, and analyzed to obtain characteristic length cumulative distributions (see Figure 7) and shape histograms [45]. The results showed that characteristic length distributions can be modelled by a power law similar to the NASA SBM and that shape distributions are independent from target thickness and impact velocity.

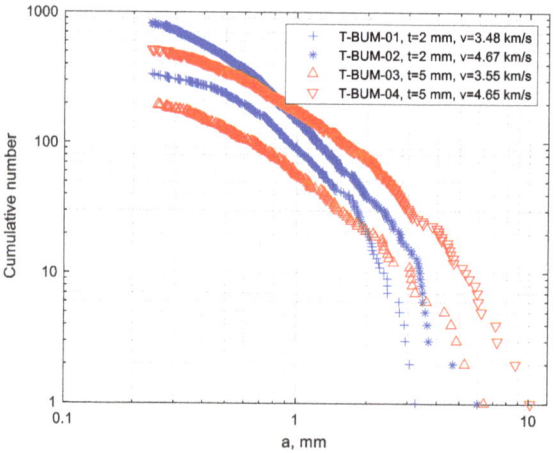

Figure 7. Fragments cumulative size distributions for the four aluminum thin plates tests [45].

A similar campaign of four tests was performed on CFRP panels (thickness of 4 mm), with spheres of 1.9 mm and 2.9 mm impacting at velocities between 3.5 and 5.1 km/s. Again, generated fragments were collected, catalogued, and analyzed to obtain size cumulative distributions (see Figure 8). It was observed that the distributions strongly differed from aluminum plates in terms of total number and shape, with a strong contribution, in size classes larger than 6 mm, of fragments generated by panel surface delamination. Results showed that the thickness to projectile diameter ratio and the impact velocity significantly affect the fragment size distributions; in particular, the number of fragments larger than 3 mm increases with these two parameters.

3.3.2. Sandwich Panels

The impact test campaign on sandwich panels consisted of four experiments: spheres of 2.9 mm in diameter were launched at velocities between 3.6 and 4.8 km/s. Three tests were carried out on aluminum-skin samples of two different thicknesses (1″ and 2″) and one experiment was performed on a CFRP-skin sandwich panel (0.5″ thick), to assess the effect of impact velocity, skin material, and target thickness [46].

Size distributions of the generated fragments can be seen in Figure 9. The skin material clearly affected the curves shape and the number of debris: samples with CFRP skins generated about one order of magnitude more fragments than aluminum ones. The results were compared with distributions from thin plates with comparable density (2 mm-thick aluminum plates, see Figure 7, and a 4 mm-thick CFRP one, see Figure 8). With respect to aluminum, the trends are comparable in terms of shape and magnitude; on the other hand, it has been observed that for CFRP, the simple plate trend is not representative of the sandwich panel, due to substantial differences in the total number of fragments and in the distribution trends.

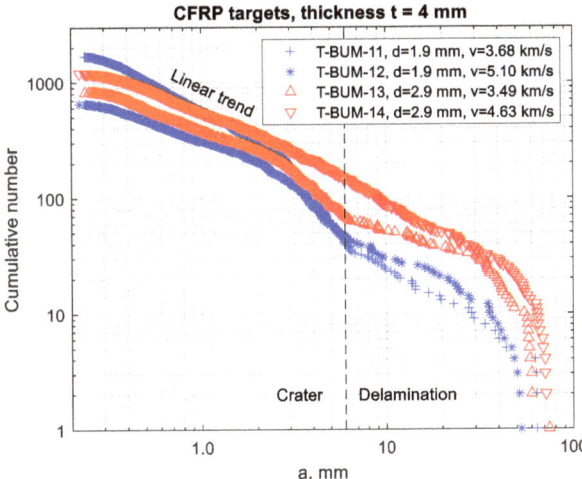

Figure 8. Fragment cumulative size distributions for the four CFRP thin plate tests.

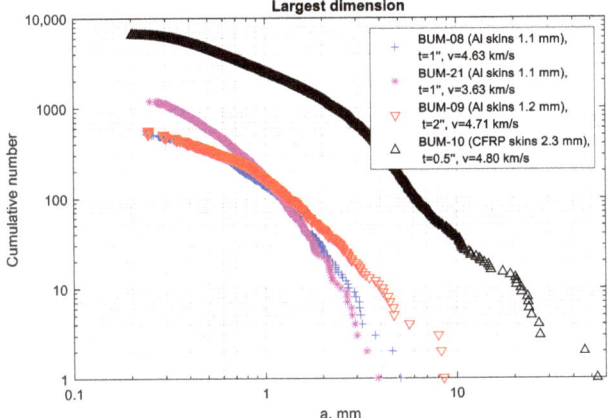

Figure 9. Fragment cumulative size distributions for the tests on sandwich panels [46].

3.3.3. Complex Mock-Up

In 2021, an impact test was performed on a $5 \times 5 \times 5$ cm^3 picosatellite mock-up with a 1.6 g nylon cylinder that hit the center of one face of the satellite at a velocity of 2.72 km/s; frames from the video of the impact can be seen in Figure 10. The mock-up included plastic elements and consumable electronic boards to simulate the different materials employed in modern spacecraft; the fraction of different materials was about 25% metals, 20% plastics, and 55% of electronic components. The collision led to the complete fragmentation of the target; the fragments were collected, weighed, and divided into size classes [47].

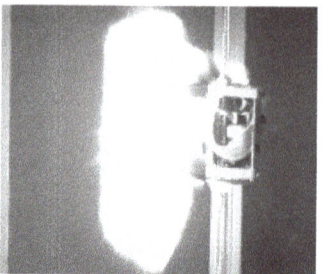

Figure 10. High-velocity camera frames from the impact video: from left to right, projectile approaching the target, collision, and impact flash [47].

The results showed that, despite the different materials employed in the mock-up, the characteristic length distribution is in line with NASA SBM prediction, even for sizes smaller than 1 cm (see Figure 11). However, the analysis of fragments' area-to-mass indicated a strong influence from fragment material and a consistent deviation from the NASA SBM prediction [47,48].

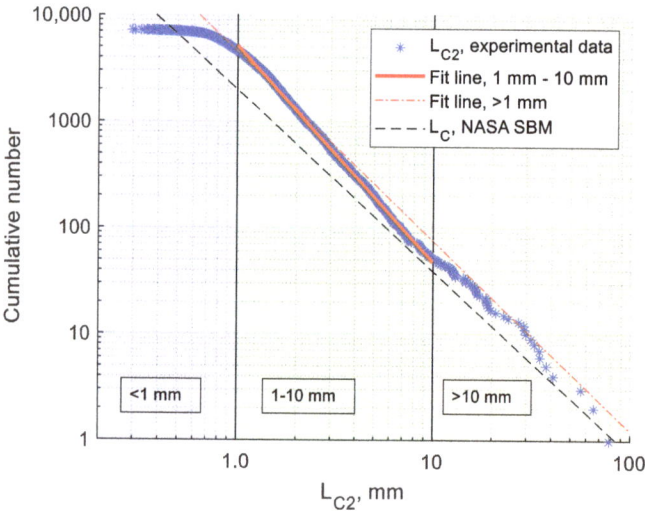

Figure 11. Comparison of experimental bi-dimensional characteristic length fitting lines for fragments larger than 1 mm (red dash-dot line) and in the range between 1 and 10 mm (solid red line) [47].

3.4. In Situ Impact Sensors

Hypervelocity impact testing can be also employed to assess the response of in-situ impact sensors for space applications [49,50]. In this context, the Space Debris group of the University of Padova was involved in the CADETS (A Calorimetric Detection System for Hypervelocity Impacts) project in collaboration with the University of Oxford and the University of Malta. This activity used Thin-Film Heat-Flux Gauges (TFHFG) to measure the local increase in shield temperature following an impact event, which is correlated with the kinetic energy of the debris. The experimental campaign was performed at the University of Padova and consisted of 4 impact tests on aluminum targets instrumented with a sensor based on TFHFG technology. Results showed contributions in sensors reading from both thermal and mechanical loads, with the latter predominant. More information can be found in [49].

4. Impact Modelling Tool

The current tools employed to simulate spacecraft catastrophic impacts and fragmentation implement empirical or semi-empirical algorithms and are based on observations of impact events and experimental data from ground tests. To date, such tools present limitations related to the novel materials employed on modern satellites (e.g., CFRP and other composites) and the evolution of manufacturing solutions; in addition, impact scenarios including glancing impacts cannot be considered in their complexity. To overcome the limitations of such solutions, the University of Padova developed the Collision Simulation Tool Solver (CSTS) in the framework of a contract awarded by the European Space Agency to CISAS and the German enterprise Etamax Space GmbH [51]. In CSTS, colliding objects are modelled with a mesh of Macroscopic Elements (MEs), representing major satellite parts, connected by structural links to form a system-level net (see Figure 12). During an impact event, the involved MEs are subjected to fragmentation, and structural damage can be transmitted through the links net; the generated fragments can affect the other MEs, creating a cascade effect representative of the object fragmentation. Based on this modelling concept, the simulator core of the CSTS is divided into three main parts: the ME Breakup Algorithm, providing fragments size, velocity, and area-to-mass distributions for a variety of spacecraft building blocks; the Structural Response Algorithm, calculating momentum transfer, energy dissipation, structural deformation, and fracture; and the Fragments Tracking Algorithm, which follows the trajectories of new debris created in the early stages of the event and detects the resulting multiple secondary impacts on other satellite parts.

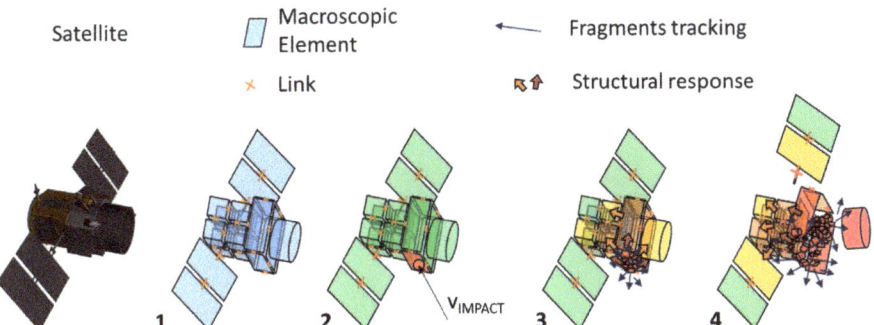

Figure 12. CSTS logic: the spacecraft is modelled as a net of macroscopic element connected by structural links (**1**); during impact (**2**), the collided ME fragments (**3**), and the start of a cascade effect involving the other MEs (**4**).

This approach allows CSTS to simulate complex collision scenarios with design details included, producing statistically accurate results. In particular, CSTS allows the reduction of the complexity of the fragmentation analysis, since it is based on a set of ME fragmentation models that are easier to develop, tune, and update through testing. It was therefore possible to validate CSTS using sub-scale test results and data available in the literature.

5. Numerical Simulation Activities

CSTS was employed to investigate a set of impact scenarios and compare the results with current fragmentation models.

5.1. Impacts with a LEO Satellite: LOFT

The first set of simulations focused on hypervelocity collisions involving the ESA LOFT spacecraft, with CubeSat impactors ranging from 1U to 48U and diverse collision scenarios. The study was conducted with the purpose of investigating the transition between sub-catastrophic and catastrophic collision while increasing the impactor's kinetic energy, and the effect of impact point and encounter configuration on the collision severity. Among the eight investigated collision scenarios (see Figure 13), the first three cases assessed LOFT fragmentation at different values of the EMR parameter. The next three cases intended to study the effect of impact point and impactor/target overlap when the EMR exceeds the accepted 40 J/g threshold for catastrophic breakup. The last two cases examined the collision of a larger body with a trajectory directed toward the denser part of the LOFT spacecraft [52].

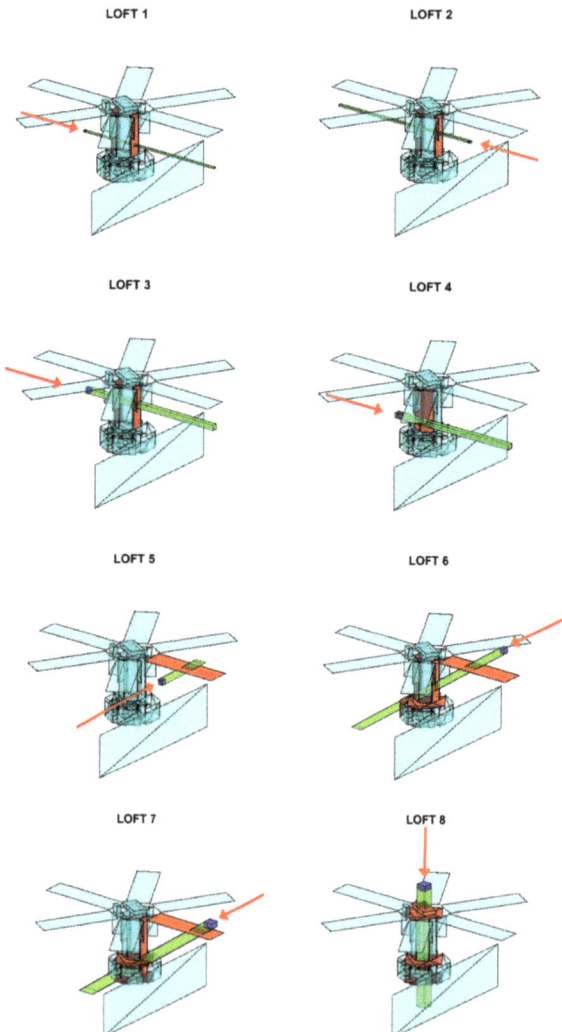

Figure 13. Collision configurations for the LOFT simulation campaign. The arrows indicate the impactors and the velocity vector, and the impactors cross-section is extruded along the velocity vector to show the nominal path through the target [52].

The simulation results indicate that the classic EMR parameter is not sufficient to model the transition between sub-catastrophic and catastrophic impact in a certain variety of collision scenarios. In fact, the same value of EMR can produce different levels of damage if the impact does not occur on the target's center of mass or the impactor's kinetic energy results from different combinations of impactor mass and velocity. Since these two points are not considered in the current version of the NASA SBM, it is also shown that the NASA model may overestimate the fragment distributions when impacts are not central or the ratio between impactor size and target size is small [52].

5.2. Impacts with GEO Objects

The second set of simulated collisions involved GEO telecommunication satellites with sizes comparable to the Intelsat spacecraft family [53]. The main goals of this work were to assess the transition from non-hypervelocity to hypervelocity regime, as low-velocity scenarios are of primary importance for GEO collisions, and to evaluate the influence of structural properties and impact point (when appendages interact first) to possibly dissipate impact energy and dampen the event. In the simulation campaign, both impactor and target had the same design, apart from a geometric scale factor equal to 1:0.535 (target and impactor mass were 3280 kg and 500 kg, respectively) and contained detailed elements, such as internal components, instrumentation, and antennas. The two impact conditions can be seen in Figure 14 and consist of direct impacts on the central bodies or involving the spacecraft appendages first (solar panels). For each configuration, three impact velocities were simulated, 0.1, 1, and 10 km/s, which led to an EMR of 0.66, 66, and 6600 kJ/kg, respectively.

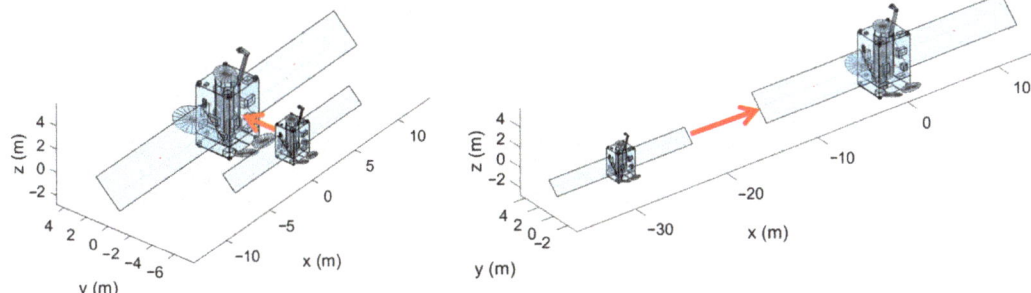

Figure 14. Simulated collision scenarios: central impacts (**left**) and collisions on the appendages (**right**).

Among the main results collected by this work and available in [53], it was found that the large satellite is never completely fragmented, despite the increase in the number of fragments with the impact velocity. In addition, for the 10 km/s scenarios, the impact configurations have a marginal effect on the total number of generated fragments and the number of fragments larger than 10 cm is higher in the collisions involving the appendages.

5.3. ENVISAT Collision Scenarios

The third set of simulations investigated potential collisions involving ENVISAT as target, see Figure 15 [54]. In this case, collision scenarios included two impactors (a 100 kg satellite and a 3-ton rocket stage) at two impact velocities (1 and 10 km/s), colliding with ENVISAT on its central body or its main appendage. Among the main results collected by this campaign of simulations, it was found that the 40 kJ/kg threshold was not representative of the transition to catastrophic impact and that the current NASA SBM significantly overestimates the number of fragments larger than 1 m. More details are reported in [54].

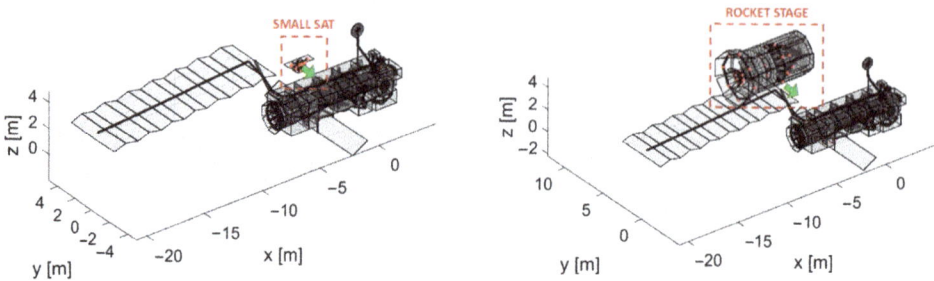

Figure 15. Simulated collision scenarios involving ENVISAT and a 100 kg small sat (**left**) or a 3-ton rocket stage (**right**).

5.4. Fragmentation Model Development

The simulation campaigns performed with the CSTS allowed us to conclude that there is room for improvement in the fragmentation models currently employed (e.g., NASA SBM) and, in particular, in the definition of the threshold for catastrophic impacts. A novel formulation should consider the impact geometry as well as the presence of appendages that might partially shield the involved bodies, at least for less energetic impacts. In this context, the Space Debris group proposed a new piecewise analytical fragmentation model, derived from simulation data, showing a good agreement with the fragment distribution trends [55].

6. Conclusions

This paper presents some recent research activities conducted by the University of Padova in the framework of Space Debris. The presence of a hypervelocity impact facility, with a high shot frequency up to ten experiments per day, makes it possible to perform experimental campaigns on a wide number of targets, from advanced shields to smart structures and sensors. With respect to larger hypervelocity impact facilities, the LGG available at the University of Padova has a maximum velocity of 5.5 km/s for projectiles with a mass of up to 100 mg; investigating higher velocities requires numerical simulations, as long as their validation can be performed within the LGG operative range. The development and installation of an upgraded LGG is under evaluation; in parallel, the formal collaboration with other laboratories which complements operational parameters will allow the enhancement of the available ranges of projectiles masses and velocities. The performed experimental activities spanned from materials and structures survivability testing to the investigation of the fragmentation processes. The results from the test campaigns were in line with data collected from other institutions; in particular, fragmentation distributions can be compared with experimental results for both simple and complex targets.

In parallel, the development of the Collision Simulation Tool Solver allowed us to study complex orbital collision scenarios with large impactors and full details included. The CSTS is therefore a powerful tool for investigating catastrophic impacts and improving current spacecraft fragmentation models. The current challenges in CSTS are the extension of the number of available macroscopic elements, to include electronic boards, solar panels, and pressurized tanks; material libraries are also still limited to the most common materials employed in the space sector.

To overcome these limitations, current experimental activities focus on material fragmentation, in the framework of commercial contracts with enterprises and institutions; numerical investigations are similarly dedicated to updating the CSTS breakup models and to upgrading the catastrophic impact threshold formulation.

Author Contributions: Conceptualization, L.O., data curation, C.G. and S.L.; writing—original draft preparation, L.O. and S.L.; writing—review and editing, C.G. and A.F.; supervision, A.F.; project administration. All authors have read and agreed to the published version of the manuscript.

Funding: The IMBEMUS project was supported by the CARIPARO foundation. Results presented in Section 3.2.3 were funded through the European Commission Horizon 2020, Framework Programme for Research and Innovation (2014–2020), under the ReDSHIFT project (grant agreement no. 687500). The CADETS project was partially funded by the Malta Council for Science and Technology. The CSTS development and the fragmentation tests were supported by the European Space Agency through contracts No. 4000119143/16/NL/BJ/zk, "Numerical simulations for spacecraft catastrophic disruption analysis", and No. 4000133656/20/D/SR, "Exploiting numerical modelling for the characterisation of collision break-ups". Part of this work is supported by the Italian Space Agency, in the framework of the ASI-INAF Agreement "Supporto alle attivita' IADC e validazione pre-operativa per SST (N. 2020-6-HH.0)".

Data Availability Statement: Not applicable.

Conflicts of Interest: The authors declare no conflict of interest.

Abbreviations

BETs	Bare Electrodynamic Tethers
BLE	Ballistic Limit Equation
CADETS	A Calorimetric Detection System for Hypervelocity Impacts
CFRP	Carbon-Fiber-Reinforced Panels
CISAS	Center for Space Activities "G. Colombo"
CSTS	Collision Simulation Tool Solver
ENVISAT	ENVIronmental SATellite
EMR	Energy-to-Mass Ratio
ESA	European Space Agency
E.T.Pack	Electrodynamic Tether technology for PAssive Consumable-less deorbit Kit
FASTT	Fragmentation Algorithms for Strategic and Theater Targets
GEO	Geosynchronous Equatorial Orbit
IMBEMUS	IMpact BEhavior of MUltifunctional materialS
JAXA	Japan Aerospace Exploration Agency
LEO	Low Earth Orbit
LGG	Light-Gas Gun
LOFT	Large Observatory For X-ray Timing
ME	Macroscopic Element
NASA	National Aeronautics and Space Administration
PEEK	PolyEther Ether Ketone
ReDSHIFT	Revolutionary Design of Spacecraft through Holistic Integration of Future Technologies
SBM	Standard Breakup Model
TFHFG	Thin-Film Heat-Flux Gauges

References

1. Karacalioglu, A.G.; Jan, S. The impact of new trends in satellite launches on the orbital debris environment. In Proceedings of the IAASS Conference Safety First, Safety for All, Melbourne, FL, USA, 18–20 May 2016; No. ARC-E-DAA-TN31699.
2. Olivieri, L.; Francesconi, A. Large constellations assessment and optimization in LEO space debris environment. *Adv. Space Res.* **2020**, *65*, 351–363. [CrossRef]
3. Lewis, H.; Radtke, J.; Rossi, A.; Beck, J.; Oswald, M.; Anderson, P.; Bastida Virgili, B.; Krag, H. Sensitivity of the space debris environment to large constellations and small satellites. *J. Br. Interplanet. Soc.* **2017**, *70*, 105–117.
4. Kessler, D.J.; Cour-Palais, B.G. Collision frequency of artificial satellites: The creation of a debris belt. *J. Geophys. Res. Space Phys.* **1978**, *83*, 2637–2646. [CrossRef]
5. Kessler, D.J.; Johnson, N.L.; Liou, J.C.; Matney, M. The Kessler syndrome: Implications to future space operations. *Adv. Astronaut. Sci.* **2010**, *137*, 2010.
6. Anselmo, L.; Rossi, A.; Pardini, C. Updated results on the long-term evolution of the space debris environment. *Adv. Space Res.* **1999**, *23*, 201–211. [CrossRef]

7. McKnight, D.; Di Pentino, F.; Knowles, S. Massive Collisions In LEO—A Catalyst To Initiate ADR. In Proceedings of the 65th International Astronautical Congress, Toronto, ON, Canada, 3 October–19 September 2014.
8. Yakovlev, M. The "IADC Space Debris Mitigation Guidelines" and supporting documents. In Proceedings of the 4th European Conference on Space Debris, Darmstadt, Germany, 18–20 April 2005; ESA Publications Division: Noordwijk, The Netherlands, 2005; Volume 587, pp. 591–597.
9. Mark, C.P.; Surekha, K. Review of active space debris removal methods. *Space Policy* **2019**, *47*, 194–206. [CrossRef]
10. Tadini, P.; Tancredi, U.; Grassi, M.; Anselmo, L.; Pardini, C.; Francesconi, A.; Branz, F.; Maggi, F.; Lavagna, M.; DeLuca, L.; et al. Active debris multi-removal mission concept based on hybrid propulsion. *Acta Astronaut.* **2014**, *103*, 26–35. [CrossRef]
11. Colombo, C.; Rossi, A.; Dalla Vedova, F.; Francesconi, A.; Bombardelli, C.; Trisolini, M.; Gonzalo, J.L.; Di Lizia, P.; Giacomuzzo, C.; Khan, S.B.; et al. Effects of passive de-orbiting through drag and solar sails and electrodynamic tethers on the space debris environment. In Proceedings of the 69th International Astronautical Congress (IAC 2018), Bermen, Germany, 1–5 October 2018; IAF: Chelsea, QC, Canada, 2018.
12. Olivieri, L.; Valmorbida, A.; Sarego, G.; Lungavia, E.; Vertuani, D.; Lorenzini, E.C. Test of tethered deorbiting of space debris. *Adv. Astronaut. Sci. Technol.* **2020**, *3*, 115–124. [CrossRef]
13. Krag, H.; Serrano, M.; Braun, V.; Kuchynka, P.; Catania, M.; Siminski, J.; Schimmerohn, M.; Marc, X.; Kuijper, D.; Shurmer, I.; et al. A 1 cm space debris impact onto the sentinel-1a solar array. *Acta Astronaut.* **2017**, *137*, 434–443. [CrossRef]
14. Kelso, T.S. *Analysis of the Iridium 33 Cosmos 2251 Collision*; NTRA: Washington, DC, USA, 2009.
15. Christiansen, E.L. Design and performance equations for advanced meteoroid and debris shields. *Int. J. Impact Eng.* **1993**, *14*, 145–156. [CrossRef]
16. Ryan, S.; Schaefer, F.; Destefanis, R.; Lambert, M. A ballistic limit equation for hypervelocity impacts on composite honeycomb sandwich panel satellite structures. *Adv. Space Res.* **2008**, *41*, 1152–1166. [CrossRef]
17. Francesconi, A.; Giacomuzzo, C.; Feltrin, F.; Antonello, A.; Savioli, L. An engineering model to describe fragments clouds propagating inside spacecraft in consequence of space debris impact on sandwich panel structures. *Acta Astronaut.* **2015**, *116*, 222–228. [CrossRef]
18. Schäfer, F.K.; Ryan, S.; Lambert, M.; Putzar, R. Ballistic limit equation for equipment placed behind satellite structure walls. *Int. J. Impact Eng.* **2008**, *35*, 1784–1791. [CrossRef]
19. Schonberg, W.P.; Martin Ratliff, J. Hypervelocity impact of a pressurized vessel: Comparison of ballistic limit equation predictions with test data and rupture limit equation development. *Acta Astronaut.* **2015**, *115*, 400–406. [CrossRef]
20. Johnson, N.L.; Krisko, P.H.; Liou, J.-C.; Anz-Meador, P.D. NASA's new breakup model of EVOLVE 4.0. *Adv. Space Res.* **2001**, *28*, 1377–1384. [CrossRef]
21. McKnight, D.; Maher, R.; Nagl, L. Fragmentation Algorithms for Strategic and Theater Targets (FASTT) Empirical Breakup Model, Ver 3.0. *Int. J. Impact Eng.* **1994**, *17*, 547–558. [CrossRef]
22. McKnight, D.; Maher, R.; Nagl, L. Refined algorithms for structural breakup due to hypervelocity impact. *Int. J. Impact Eng.* **1995**, *17*, 547–558. [CrossRef]
23. Sorge, M.E. Satellite fragmentation modeling with IMPACT. In Proceedings of the AIAA/AAS Astrodynamics Specialist Conference and Exhibit, Honolulu, HI, USA, 18–21 August 2008.
24. Sorge, M.E.; Mains, D.L. IMPACT fragmentation model developments. *Acta Astronaut.* **2016**, *126*, 40–46. [CrossRef]
25. Murray, J.; Cowarding, J.; Liou, J.-C.; Sorge, M.; Fitz-Coy, N.; Huynh, T. Analysis of the DebriSat fragments and comparison to the NASA standard satellite breakup model. In Proceedings of the International Orbital Debris Conference (IOC), Sugar Land, TX, USA, 4–7 December 2019; No. JSC-E-DAA-TN73918.
26. Hanada, T.; Liou, J.C.; Nakajima, T.; Stansbery, E. Outcome of recent satellite impact experiments. *Adv. Space Res.* **2009**, *44*, 558–567. [CrossRef]
27. Abdulhamid, H.; Bouat, D.; Colle, A.; Lafite, J.; Limido, J.; Midani, I.; Omaly, P. On-ground HVI on a nanosatellite. impact test, fragments recovery and characterization, impact simulations. In Proceedings of the 8th ESA Space Debris Conference, Darmstadt, Germany, 20 April 2021.
28. Rivero, M.; Shiotani, B.; Carrasquilla, M.; Fitz-Coy, N.; Liou, J.C.; Sorge, M.; Huynh, T.; Opiela, J.; Krisko, P.; Cowardin, H. DebriSat fragment characterization system and processing status. In Proceedings of the International Astronautical Congress, Guadalajara, Mexico, 26–30 September 2016; No. JSC-CN-37403-1.
29. Angrilli, F.; Pavarin, D.; De Cecco, M.; Francesconi, A. Impact facility based upon high frequency two-stage light-gas gun. *Acta Astronaut.* **2003**, *53*, 185–189. [CrossRef]
30. Pavarin, D.; Francesconi, A. Improvement of the CISAS high-shot-frequency light-gas gun. *Int. J. Impact Eng.* **2003**, *29*, 549–562. [CrossRef]
31. Pavarin, D.; Francesconi, A.; Angrilli, F. A system to damp the free piston oscillations in a two-stage light-gas gun used for hypervelocity impact experiments. *Rev. Sci. Instrum.* **2004**, *75*, 245–252. [CrossRef]
32. Francesconi, A.; Pavarin, D.; Bettella, A.; Angrilli, F. A special design condition to increase the performance of two-stage light-gas guns. *Int. J. Impact Eng.* **2008**, *35*, 1510–1515. [CrossRef]
33. Cosmo, M.L.; Enrico, C.L. *Tethers in Space Handbook*; No. NASA/CR-97-206807; Open Library: Washington, DC, USA; Springfield, VA, USA, 1997.

34. Anselmo, L.; Carmen, P. The survivability of space tether systems in orbit around the earth. *Acta Astronaut.* **2005**, *56*, 391–396. [CrossRef]
35. Francesconi, A.; Giacomuzzo, C.; Bettiol, L.; Lorenzini, E. A new Ballistic Limit Equation for thin tape tethers. *Acta Astronaut.* **2016**, *129*, 325–334. [CrossRef]
36. Khan, S.B.; Francesconi, A.; Giacomuzzo, C.; Lorenzini, E. Survivability to orbital debris of tape tethers for end-of-life spacecraft de-orbiting. *Aerosp. Sci. Technol.* **2016**, *52*, 167–172. [CrossRef]
37. Valmorbida, A.; Olivieri, L.; Brunello, A.; Sarego, G.; Sánchez-Arriaga, G.; Lorenzini, E. Validation of enabling technologies for deorbiting devices based on electrodynamic tethers. *Acta Astronaut.* **2022**, *198*, 707–719. [CrossRef]
38. Francesconi, A.; Giacomuzzo, C.; Kibe, S.; Nagao, Y.; Higashide, M. Effects of high-speed impacts on CFRP plates for space applications. *Adv. Space Res.* **2012**, *50*, 539–548. [CrossRef]
39. Francesconi, A.; Giacomuzzo, C.; Grande, A.; Mudric, T.; Zaccariotto, M.; Etemadi, E.; Di Landro, L.; Galvanetto, U. Comparison of self-healing ionomer to aluminium-alloy bumpers for protecting spacecraft equipment from space debris impacts. *Adv. Space Res.* **2013**, *51*, 930–940. [CrossRef]
40. Grande, A.M.; Coppi, S.; Di Landro, L.; Sala, G.; Giacomuzzo, C.; Francesconi, A.; Rahman, M.A. An experimental study of the self-healing behavior of ionomeric systems under ballistic impact tests. In *Behavior and Mechanics of Multifunctional Materials and Composites 2012*; SPIE: Bellingham, WA, USA, 2012; Volume 8342.
41. Mudric, T.; Giacomuzzo, C.; Francesconi, A.; Galvanetto, U. Experimental investigation of the ballistic response of composite panels coupled with a self-healing polymeric layer. *J. Aerosp. Eng.* **2016**, *29*, 04016047. [CrossRef]
42. Rossi, A.; Alessi, E.M.; Schettino, G.; Beck, J.; Holbrough, I.; Schleutker, T.; Letterio, F.; Vicario de Miguel, G.; Becedas Rodriguez, J.; Dalla Vedova, F.; et al. The H2020 ReDSHIFT project: A successful European effort towards space debris mitigation. In Proceedings of the 70th International Astronautical Congress (IAC 2019), Washington, DC, USA, 21–25 October 2019.
43. Olivieri, L.; Giacomuzzo, C.; Francesconi, A.; Stokes, H.; Rossi, A. Experimental characterization of multi-layer 3D-printed shields for microsatellites. *J. Space Saf. Eng.* **2020**, *7*, 125–136. [CrossRef]
44. Dimare, L.; Francesconi, A.; Giacomuzzo, C.; Cicalò, S.; Rossi, A.; Olivieri, L.; Sarego, G.; Guerra, F.; Lemmens, S.; Braun, V. Advances in the characterisation of collision break-ups by means of numerical modelling. In Proceedings of the 72th International Astronautical Congress (IAC 2021), Dubai, United Arab Emirates, 25–29 October 2021.
45. Olivieri, L.; Cinzia, G.; Alessandro, F. Experimental fragments distributions for thin aluminium plates subjected to hypervelocity impacts. *Int. J. Impact Eng.* **2022**, *170*, 104351. [CrossRef]
46. Olivieri, L.; Cinzia, G.; Alessandro, F. Impact fragments from honeycomb sandwich panels. In Proceedings of the 73rd International Astronautical Congress, Paris, France, 18–22 September 2022.
47. Olivieri, L.; Smocovich, P.A.; Giacomuzzo, C.; Francesconi, A. Characterization of the fragments generated by a Picosatellite impact experiment. *Int. J. Impact Eng.* **2022**, *168*, 104313. [CrossRef]
48. Olivieri, L.; Cinzia, G.; Alessandro, F. Analysis of fragments larger than 2 mm generated by a picosatellite fragmentation experiment. *Acta Astronaut.* **2023**, *204*, 418–424. [CrossRef]
49. Barilaro, L.; Falsetti, C.; Olivieri, L.; Giacomuzzo, C.; Francesconi, A.; Beard, P.; Camilleri, R. A conceptual study to characterize properties of space debris from hypervelocity impacts through Thin Film Heat Flux Gauges. In Proceedings of the IEEE MetroAerospace, Naples, Italy, 23–25 June 2021.
50. Colombo, C.; Trisolini, M.; Scala, F.; Brenna, M.P.; Gonzalo Gòmez, J.L.; Antonetti, S.; Di Tolle, F.; Redaelli, R.; Lisi, F.; Marrocchi, L.; et al. Cube mission: The environmental CubeSat. In Proceedings of the 8th European Conference on Space Debris, ESA/ESOC, Darmstadt, Germany, 20 April 2021; ESA: Paris, France, 2021.
51. Francesconi, A.; Giacomuzzo, C.; Olivieri, L.; Sarego, G.; Duzzi, M.; Feltrin, F.; Valmorbida, A.; Bunte, K.D.; Deshmukh, M.; Farahvashi, E.; et al. CST: A new semi-empirical tool for simulating spacecraft collisions in orbit. *Acta Astronaut.* **2019**, *160*, 195–205. [CrossRef]
52. Francesconi, A. Examination of satellite collision scenarios spanning low to hypervelocity encounters using semi-empirical models. In Proceedings of the 70th International Astronautical Congress (IAC), Washington, DC, USA, 21–25 October 2019.
53. Francesconi, A.; Giacomuzzo, C.; Olivieri, L.; Sarego, G.; Valmorbida, A.; Duzzi, M.; Bunte, K.D.; Farahvashi, E.; Cardone, T.; de Wilde, D. Numerical simulations of hypervelocity collisions scenarios against a large satellite. *Int. J. Impact Eng.* **2022**, *162*, 104130. [CrossRef]
54. Olivieri, L.; Giacomuzzo, C.; Duran-Jimenez, C.; Francesconi, A.; Colombo, C. Fragments Distribution Prediction for ENVISAT Catastrophic Fragmentation. In Proceedings of the 8th European Conference on Space Debris, ESA/ESOC, Darmstadt, Germany, 20 April 2021; ESA: Paris, France, 2021.
55. Olivieri, L.; Giacomuzzo, C.; Francesconi, A. Simulations of satellites mock-up fragmentation. *Acta Astronaut.* **2023**, *206*, 233–242. [CrossRef]

Disclaimer/Publisher's Note: The statements, opinions and data contained in all publications are solely those of the individual author(s) and contributor(s) and not of MDPI and/or the editor(s). MDPI and/or the editor(s) disclaim responsibility for any injury to people or property resulting from any ideas, methods, instructions or products referred to in the content.

Article

Design of the Sabot-Stopping System for a Single-Stage Light-Gas Gun for High-Velocity Impacts

Leonardo Barilaro [1,*], Mark Wylie [2] and Theeba Shafeeg [2]

[1] Department of Aviation, The Malta College of Arts, Science & Technology, Triq Kordin, PLA 9032 Paola, Malta
[2] Department of Aerospace, Mechanical and Electronic Engineering, South East Technological University (SETU), Carlow Campus, Kilkenny Rd, Moanacurragh, R93 V960 Carlow, Ireland; mark.wylie@setu.ie (M.W.)
* Correspondence: leonardo.barilaro@mcast.edu.mt

Abstract: Collisions of space debris and micrometeorites with spacecraft represent an existential hazard for human activities in near-Earth orbits. Currently, guidelines, policies, and best practices are encouraged to help mitigate further propagation of this space debris field from redundant spacecraft and satellites. However, the existing space debris field is an environment that still poses a great threat and requires the design of contingency and fail-safe systems for new spacecraft. In this context, both the monitoring and tracking of space debris impact paths, along with knowledge of spacecraft design features that can withstand such impacts, are essential. Regarding the latter, terrestrial test facilities allow for replicating of space debris collisions in a safe and controlled laboratory environment. In particular, light-gas guns allow launching impactors at speeds in the high-velocity and hypervelocity ranges. The data acquired from these tests can be employed to validate in-orbit observations and structural simulations and to verify spacecraft components' survivability. Typically, projectiles are launched and protected using a sabot system. This assembly, known as a launch package, is fired towards a sabot-stopping system. The sabot separates from the rest of the launch package, to avoid target contamination, and allows the projectile to travel towards the target through an opening in the assembly. The response and survivability of the sabot-stopping system, along with the transmission of the forces to the light-gas gun structure and prevention of target contamination, is an important design feature of these test apparatuses. In the framework of the development of Malta's first high-velocity impact facility, particular attention was dedicated to this topic: in this paper, the description of a novel sabot-stopping system is provided. The system described in this research is mechanically decoupled from the interaction with the impact chamber and the light-gas gun pump tube; this solution avoids damage in case of failures and allows easier operations during the pre- and post-test phases.

Keywords: space debris; sabot-stopping system; high-velocity impacts

1. Introduction

Space debris refers to man-made objects that have been left in orbit around the Earth, such as old satellites, rocket stages, and debris from past space missions. They can range in size from tiny fragments to large, multi-ton objects. The artificial space debris materials can be divided into seven types, which are polymers, non-metal debris, metals and their alloys, oxides, sulphides and their analogs, halides, and carbides. However, aluminium alloys are the most common materials that can lead to the creation of space junk. On the other hand, micrometeorites are tiny natural particles that come from comets, asteroids, and other celestial bodies. They can be formed of a range of substances, such as silicates, iron, and carbonaceous minerals, and are typically less than 1 mm in size. Due to the great speeds at which they travel, even small particles can cause significant damage upon impact, representing a risk to spacecraft and satellites in orbit around the Earth for both space debris and micrometeorites. Therefore, it is important to track and monitor these objects to help protect space assets [1].

The potential for polluting the low-Earth environment with space debris became evident shortly after the first decade of space exploration [2]; more recent investigations highlighted the continuous degradation of the debris environment [3]. In addition, the constant increase in the launch of small satellites [4] as well as the high number of vehicles for large satellite constellations pose an additional risk to the access to low-Earth orbits (LEO) [5–7]. Collisions with debris (e.g., [8]) or with uncontrolled large bodies (e.g., [9]) have already affected operational satellites. Unless the most hazardous objects are removed from LEO orbit [10] and all satellites are provided with end-of-life disposal systems [11–14], only policies and regulations can currently mitigate a further environment degradation and limit the hazards for operative satellites [15,16].

In this context, developing and updating debris population models [17] as well as analysing the vulnerability of spacecraft architectures and components [18,19] is essential to finding solutions for reducing the degradation of the debris environment. In addition, ground test activities on large [20–22] and small [23,24] spacecraft mock-ups, CubeSat [25] and Picosat [26,27] models, as well as the development of numerical simulation tools [28–32] through the utilization of ground test data and space breakup event observations are fundamental steps in understanding the generation of space debris formation [25].

On these considerations, the utilization of test facilities [33] that allow replicating collisions in a safe and controlled laboratory environment is still relevant today, and the development of this type of facility is a result of the IADC guidelines for space debris mitigation [25,34]. In particular, light-gas guns (LGG) allow launching impactors at speeds in the high-velocity and the hypervelocity ranges [35–37]. In this context, the Malta College of Arts, Science & Technology (MCAST) is developing a novel LGG facility [38] with operational ranges that will allow testing velocities typical of GEO impacts. In particular, system designs from similar facilities [39,40] have been adopted to reduce the maintenance requirements of such a facility and increase the number of tests that can be performed in a single day.

In the development of a hypervelocity impact facility, significant attention is focused on several crucial components, including the launch package and the sabot-stopping system. The launch package serves as a protection for the projectile during acceleration, while the sabot-stopping system is responsible for halting the sabot. Typically, launch-gas guns (LGGs) employ an expandable sabot system to house the projectile. Subsequently, after the acceleration phase, the sabot disengages from the projectile, fragmenting to prevent target contamination. However, this approach can cause damage to the sabot-stopping system, necessitating either complete replacement or time-consuming maintenance [2]. This research paper tackles this issue by presenting an innovative solution: a reusable and modular design for the sabot-stopping system. The objective is to develop a design that ensures operational safety, requires less maintenance compared to conventional LGG designs, and minimizes the transmission of impact forces to the LGG structure. The subsequent sections provide an overview of MCAST's LGG and a description of the sabot-stopping system, followed by an explanation of the advantages offered by this solution.

2. Single-Stage LGG Overview

In this section, an overview of the single-stage LGG being developed for the MCAST Impacts Facility is presented; in addition, the preliminary layout of the facility is described. The design of the single-stage LGG takes into account the following objectives:

- To establish a high-velocity impact test facility that can conduct testing for the aviation and space industries. The selected range of projectile velocities is suitable for testing impacts on aircraft parts as well as simulating low-speed impacts, like those observed in GEO orbit [41]. The LGG aims at replicating mainly impacts due to metal impactors, mainly aluminium alloys [42]. Its modular setup can also be adapted to other types of projectiles simulating space debris, such as plastics and silica materials.
- To set up a laboratory capable of working in conjunction with other European hypervelocity research centres, such as the one at the University of Padova [40]. The Centre

of Studies and Activities for Space "Giuseppe Colombo" (CISAS) at the University of Padova hosts a world-class experimental facility for hypervelocity impacts (HVI), utilizing a two-stage LGG system. This two-stage LGG is capable of launching projectiles weighing up to approximately 100 mg at velocities of up to 5.5 km/s. Lower velocities of up to 2.5 km/s can be reached for impactors that are heavier (up to 1 g). The MCAST single-stage LGG facility, which has been developed to accommodate higher launch package weights ranging from 15 to 40 g and enable testing at lower speeds, complements this by expanding the operational range of test campaigns. The Malta facility also has a larger impact testing chamber than the one at CISAS, providing greater experimental versatility. This increased capability makes it easier to conduct comprehensive studies on novel materials that are suitable for use in aviation and space applications.

- Establishing an LGG facility is to achieve a cost-effective experimental setup that delivers high performance and high frequency while minimizing installation and maintenance expenses.

In the following subsection, the main layout will be described.

LGG Layout

The investigation of gas dynamic aspects for a new single-stage LGG for the MCAST Hypervelocity Impacts Facility is the first step for the definition of the layout. One of the most important parameters is the projectile's velocity when it hits the target [43], with a focus on the effect of initial loading parameters on the projectile's velocity as it exits the launch tube [44]. The projectile's velocity is influenced by initial factors such as gas pressure and compressibility. According to the theory, there is a nonlinear relationship between the initial pressure and the velocity as the gas leaves the barrel. Consequently, a dedicated model in Matlab™ Simulink™ was developed to accurately simulate the performance of the LGG. The data obtained from the model serve as driver inputs for designing the first-stage reservoir, determining the pump tube length, and selecting the appropriate launch package.

Figure 1 represents the key parts of the single-stage LGG. The launch package, which consists of the protective sabot and the projectile, is accommodated in the pump tube downward and is accelerated with the high-pressure gas from the first stage. The first-stage reservoir contains an inert gas, such as helium (He), under high pressure. After the acceleration phase, a sabot-stopping system enables the projectile to be detached from the sabot; the mechanism of this system is described in Section 3. Two laser blades in the launch tube monitor the projectile's trajectory toward the target while measuring its velocity; the projectile finally impacts the target in the impact chamber.

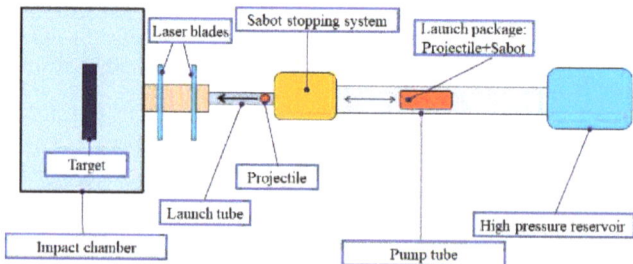

Figure 1. Schematic of a single-stage LGG.

A range of pressure values have been considered for the reservoir. The launch package of 40 g at 150 bar allows for velocities up to ≈500 m/s to be achieved using a 3 m long pump tube, as shown in Figure 2.

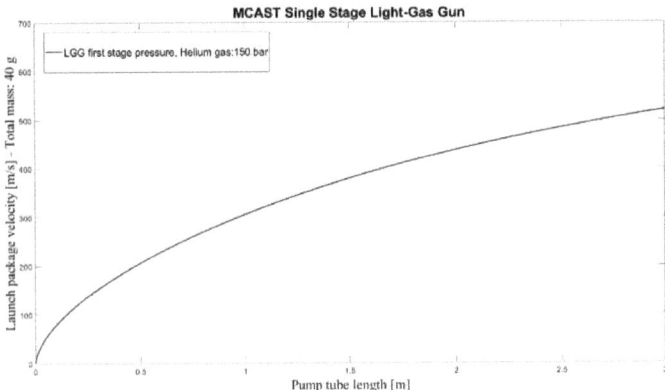

Figure 2. Single-stage LGG performances at 150 bar.

3. Sabot-Stopping System Conceptual Design

In the development of an LGG system, the structural integrity of the sabot-stopping system is critical to the repeatability and accuracy of the test apparatus [45]. The sudden deceleration, subsequent high forces, and potential for misalignment can damage and fragment the sabot-stopping system, leading to target contamination. An adequate design that will not yield or partially fail due to high strain deformation is required for the characterisation of this system, specifically input/output parameters with the sabot. While there are many serviceable elements in an LGG design, the sabot-stopping system is still the most susceptible element and may require replacement after a few impacts, even under normal operation, causing significant delays to the test campaign as the system may need calibration and characterisation after each replacement. This project proposes the design of a reusable sabot-stopping system (>20 shots) to reduce the likelihood of this issue.

The development of this concept led to the design of a technical solution where the sabot does not expand and break away, since the speed range and projectile masses involved allow for a simpler solution. It is proposed that the sabot can consist of a hollow cylinder closed at the bottom end, from which the projectile comes out after the sabot is stopped. The design of this sabot will be defined at a later stage of the described work.

The concept of the system is schematically shown in Figure 3, which also demonstrates the dissipation mechanism principle for the impact force. A tentative configuration of the key elements is shown in Figure 4. The holder for the sacrificial impact tube assembly, which consists of an impact plate with a rubber disc surrounding it and a rubber cylinder behind it, is part of the sabot system. After the shot, the primary component in interacting directly with the sabot is the impact tube. Between the impact chamber and the LGG pump tube, the holder is inserted into a supporting frame with flanges that allow for different mounting options. Several tests can be completed in a single day due to the unique design of the single-stage LGG, whose components are not destroyed or compromised during shot operations. The novel sabot-stopping system proposed here is also designed to mechanically decouple the target from the sabot-stopping system, overcoming the transmission of unwanted vibrations to the target.

This leads to more accurate and reliable results, making the proposed system highly suitable for use in LGG systems.

Additionally, the suggested sabot-stopping system enables high-velocity projectile testing while minimizing system perturbations and reducing calibration and characterization time, thereby enhancing testing efficiency. Furthermore, by eliminating the risk of target contamination, it improves the accuracy of the testing process, making it applicable in various domains [46,47]. The proposed system design carries significant implications for the progress of LGG systems, and it is expected to contribute significantly to the advancement of this field.

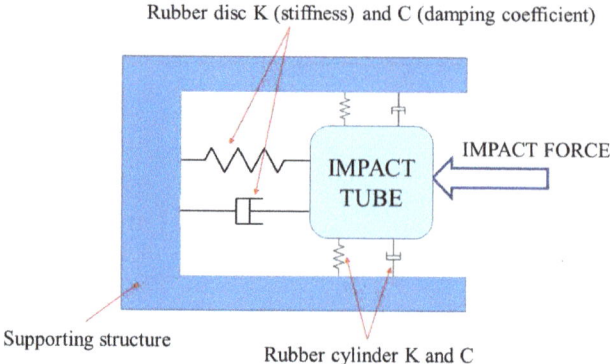

Figure 3. Schematic of the sabot-stopping system concept.

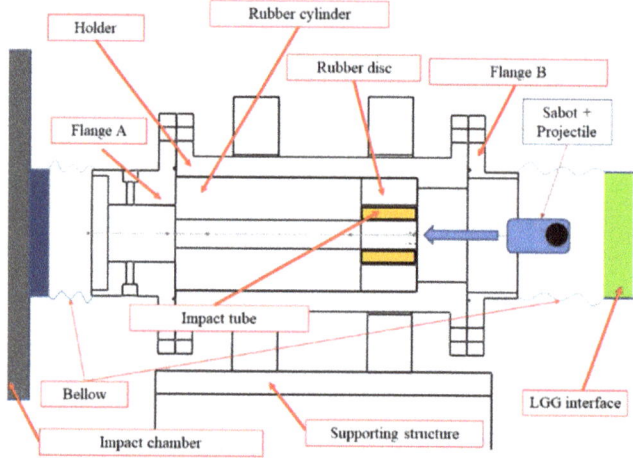

Figure 4. Schematic of the sabot-stopping system.

3.1. System Overview

The conceptual overview of the system is described here. The launch package hits the sabot-stopping system and delivers a force to the impact tube, which is dissipated via the rubber cylinder, stopping the sabot and allowing the projectile to continue its path to the target. The LGG supporting structure is protected by the transferred momentum using the properties of stiffness and damping of a rubber disc and cylinder. Radial deformations are reduced by the disc design that surrounds the tube, while axial deformations are kept to a minimum by the cylinder. By adopting this approach, it is possible to protect the holder, which is held up by the support structure and restrained by the two flanges. This structure has the advantage of allowing for the replacement and testing of a variety of impact tube types without adding to the system's complexity [48].

The detailed study of this system is divided into three phases.

A mass-spring-damper Simulink^{TM} model has first been implemented to evaluate the system's reaction force to an impact; the next phase involves doing a finite element analysis (FEA) to evaluate the behaviour of the LGG supporting structure.

The third phase involves the structural FEA of the rubber cylinder (partial). Figure 5 presents a general overview of the sabot-stopping system, while Figure 6 shows the sabot-stopping system exploded view, highlighting the main components, in particular the impact tube, rubber disc, holder, and supporting structure.

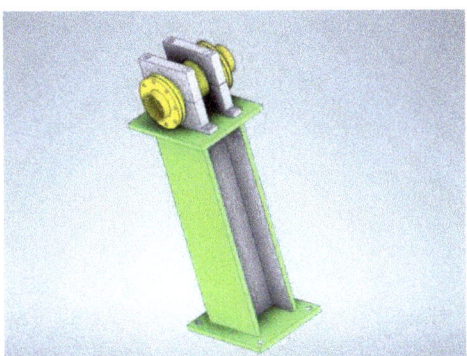

Figure 5. Overview of the sabot-stopping system.

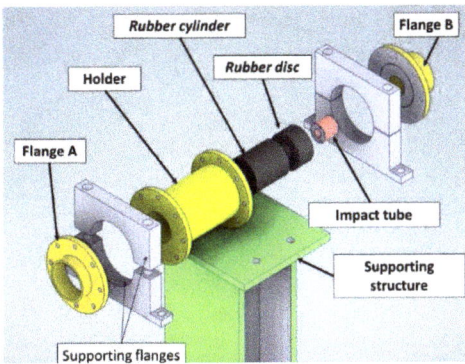

Figure 6. Sabot-stopping system exploded view.

3.2. System Design

The design of this system is based on results from the parametric dynamic model investigation of the system developed in Simulink™, the structure FEA of the LGG supporting structure, and the structural FEA of the rubber cylinder structure (partial) from the sabot-stopping system.

The first part of the design developed a mass-spring-damper Simulink™ model to assess the reaction force of the system to the impact.

The highest load condition case has been considered in order to have a conservative evaluation. The Simulink model has the following set:

- Gas: Helium at 150 bar,
- Launch package mass: m = 40 g,
- All the launch package is supposed to impact the sabot stopper (worst-case scenario).

Knowing the velocity of the launch package allows for the calculation of the transferred momentum, and following this, the reaction forces of the structure due to the impact. The first evaluations, using different time interval dt, assessed a range of this force between $F = 1 \times 10^6$ N and $F = 1 \times 10^7$ N.

The results of this first phase allow for the definition of a preliminary design of the sacrificial tube, made of Fe-310 steel, to minimise the production costs (Figure 7).

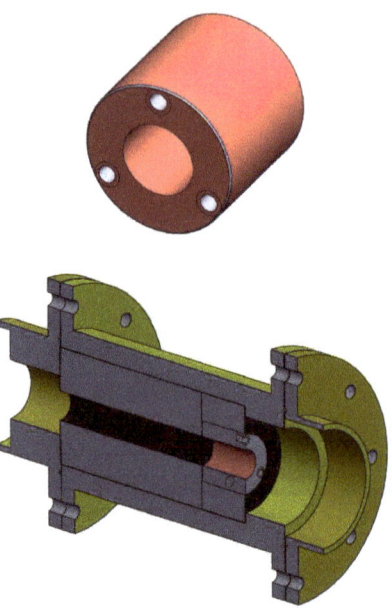

Figure 7. Impact tube (**top**) and sabot-stopping system (**bottom**).

The preliminary structural FEA to assess the behaviour of the LGG structure has been performed with the following conservative assumptions:

- All the impact forces are transmitted to the supporting structure,
- Fixed constrains are used, even though in the real situation, the other components will help in the damping effect after the shot.

The results show, Figure 8, that in this worst case, the structure's deformation is within reasonable limits (approximately 7 mm).

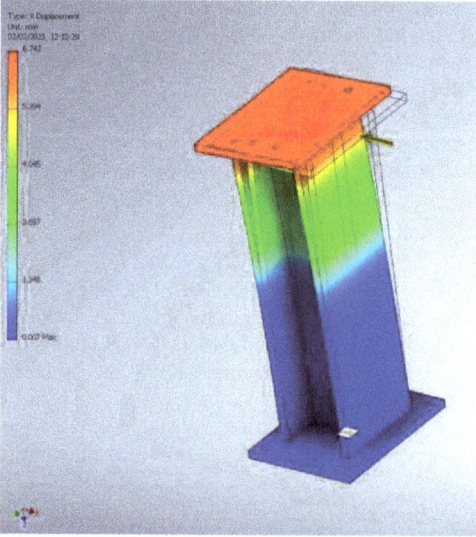

Figure 8. Sabot stopper supporting structure's total deformation.

This first set of results have been used as a baseline for the design procedure described in the following subsection. This details the third phase, which is the structural FEA of the rubber cylinder damping tube.

3.3. Energy Damping System Design

This section will describe a set of simulations carried out on the most critical energy damping component, i.e., the rubber cylinder. In an effort to understand more about the deformation and the stresses in the energy damper rubber cylinder, a non-linear dynamic hyper-elastic FEM of a 20 mm long partial section was created and shown in Figure 9. In these preliminary studies, only a section of the rubber cylinder was analysed to reduce computational resources via high element count.

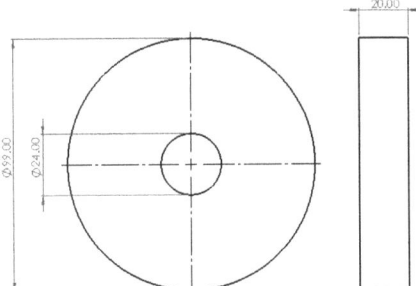

Figure 9. Rubber cylinder (20 mm section).

The main goal of these studies is to help specify the appropriate rubber material that can withstand the impact of the sabot with projectile for multiple consecutive impacts. The predicted deformation can also help with designing a key component that will not deform such that it will be placed in the path of the projectile and cause contamination of the target impact. The dimensions chosen are a first iteration to study the behaviour of the component; the results will be used to guide the final design.

For these initial studies, 5-parameter Mooney Rivlin constants ($C10 = -0.55$ MPa, $C01 = 0.7$ MPa, $C20 = 1.7$ MPa, $C11 = 2.5$ MPa, $C02 = -0.9$ MPa) were adopted and taken from the ANSYS materials library. A mass density of 1000 kg/m^3 and Poisson's ratio of 0.49 was assigned and amounted to a rubber cylinder mass of 145 g. Rayleigh damping constants for the mass (α) and stiffness (β) matrix were taken as 1.26 and 8.69×10^{-6}, respectively.

The sabot and projectile, having a mass of 40 g, will make contact with the impact tube, transferring the full momentum to the energy damping system (rubber cylinder). This is a conservative assumption. It is also assumed that the impact tube and the rubber cylinder will share the same impact surface dimensions. The velocity of this assembly is taken as 250 m/s and amounts to 1250 J. Equating strain energy to work and using an estimated travel of 0.003 m, the impact force is taken as 416.6 kN.

The rubber disk is held in a cylindrical metal tube holder and is pressed stationary against the exiting wall. The transfer of momentum upon impact between the sabot and rubber disk causes the rubber disk to compress along its longitudinal axis; but also, through conservation of volume, a reduction of the internal rubber tube diameter occurs. The full surface contact is assumed between the sabot and the rubber disk.

In the FEM, the impact force is applied to all nodes on face 1 and acting in the direction along the longitudinal axis, while the opposite end of the disk (face 2) is fixed in all D.O.F., representing the fixed exiting wall surface. Nodes on the surface of the outer cylinder wall are fixed in all D.O.F. of the cylinders longitudinal axis, i.e., the external diameter is fixed, representing the metal tube wall holder.

Figure 10 shows the model boundary conditions. The duration of the solution time is based on the period, i.e., 1/eigenfrequency. This period was then scaled by a factor of three to ensure capture of three full waves. A modal study predicted the first significant mode

(70% mass participation) at 533 Hz, resulting in a solution time of 5.62 ms. The time-step was then estimated by dividing the maximum element dimension (4 mm) in the direction of the wave propagation divided by the speed of sound in rubber, taken as 39 m/s, and this amounted to a time-step of 0.1 ms. This technique is taken from the literature [49].

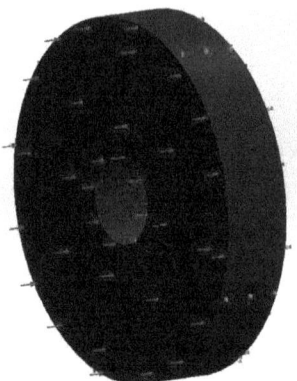

Figure 10. Rubber cylinder—FEM with boundary conditions.

Mesh density sensitivity was considered for each model. Element size, element growth ratio, and model discretization of the curved surfaces were adjusted and inspected for the maximum element aspect ratio. The meshing of the model employed blended curvature-based solid tetrahedral 3D elements with 16 Jacobian points, resulting in approximately 160 k elements and 227 k nodes, with a maximum element size of 2 mm. An element growth ratio of 1.4 was utilized, and the maximum aspect ratio was 4; 99.8% of elements remained below 3. Refer to Figure 11 for a view of the meshed model.

Figure 11. Rubber cylinder—Meshed FEM.

4. Results

The study leads to four major findings, i.e., a design layout for the LGG, a novel sabot-stopping system, a structural FEA revealing deflection of the sabot stand under a specified force, and the development of a FEM for the rubber cylinder component.

Regarding the first point, the design layout for the single-stage LGG at MCAST is presented, with a chosen configuration featuring a 3 m pump tube and launch package of 40 g, able to reach a maximum velocity of ≈500 m/s. This approach was selected to have a complementary operative range with other facilities hosting a two-stage LGG, similar to the LGG at the CISAS of the University of Padova, which is able to shoot at higher speeds but with smaller projectile masses.

A novel sabot-stopping system is also introduced. It is unique from standard LGG systems that absorb a sabot which breaks apart into multiple sections. Instead, the proposed system allows for a novel design that can absorb and dissipate the energy of a solid sabot that can be machined from a single cylinder and thus, allows for multiple shots using the same sabot-stopping setup, reducing cost and test set-up time.

A structural FEA of the sabot-stopping system stand, the structure that mounts the system to the concrete ground, was carried out. It was found that the sabot stand has a maximum deflection of 7 mm when hit with a 1×10^7 N force.

A FEM was developed for one of the energy damping components of the system, the rubber cylinder.

The FEM was used to estimate the maximum von Mises stress predicted for the rubber cylinder, which was found to be 2.46 MPa (Figure 12). This value is lower than the typical yield strength for rubber (10 MPa) and is within acceptable limits. It is important to note, however, that the stress distribution is not uniform across the rubber cylinder. In fact, the maximum stress occurs at the inner hole on the opposite face to the impact. This could potentially cause an issue if the projectile clearance with the impactor is less than 2 mm, as the maximum displacement predicted (1.96 mm) occurs at the inner hole (Figure 13). This results in a conical decrease of the diameter, which could lead to a collision or interference of the projectile with the sabot-stopping system. To illustrate this issue, Figure 14 has been included, which shows the potential for collision or interference. If such an event were to occur, it could cause the failure of the region in the impactor tube, which could result in contamination of the projectile and/or the target. It is therefore important to carefully consider the displacement relating to the reduction of the length of the impactor tube, which amounted to 0.8 mm.

The study focused on analysing the von Mises stress distribution resulting from a single impact between the sabot and the rubber stopper. However, it is important to note that the study, at this stage, did not include an evaluation of fatigue (i.e., performance requirements for the required number of tests), damage tolerance, or fail-safe considerations, all of which are crucial for design and safety purposes.

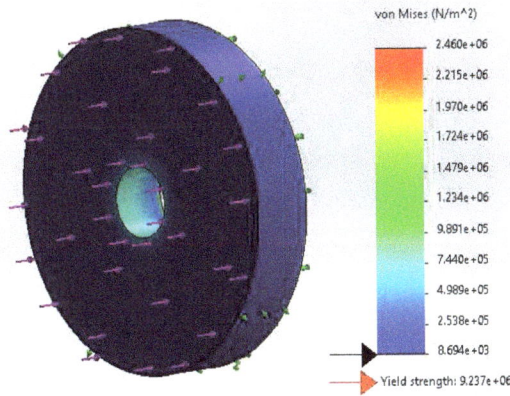

Figure 12. Resultant von Mises stress.

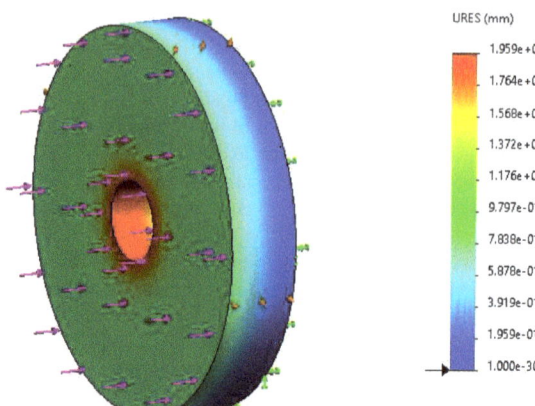

Figure 13. Resultant displacement.

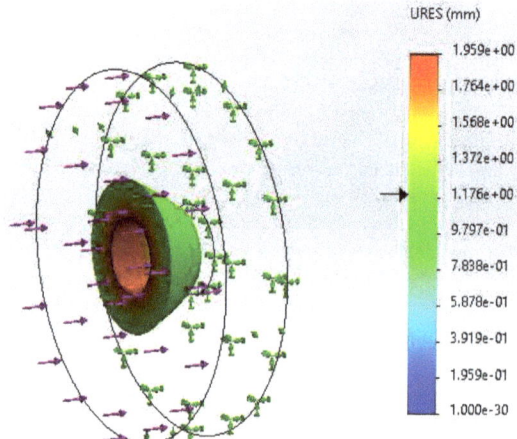

Figure 14. Resultant conical displacement profile (iso-clipped at 1.2 mm).

5. Conclusions

This paper presents an overview of the development of Malta's first high-velocity impact facility, with focus on the design process of a novel sabot-stopping system. The system described is mechanically decoupled in the interaction with the impact chamber and the light-gas gun pump tube; this solution avoids damage in the event of failures and allows easier operation during the pre- and post-test phases.

The structural FEM of the rubber cylinder after impact with the sabot predicts a displacement that results in a decrease of the rubber stopper internal diameter of ≈2 mm. Elastomeric material selection, such as EPDM or nitrile, and specification according to the designated standard will be used in future FEMs. The simulations will also be extended to a velocity range up to 500 m/s.

Material characterisation using a prototype is essential for non-linear behaviour estimation and FEM validation.

Results confirm the feasibility of the proposed design and identify the main parameters determining the performance of the full system. Based on the preliminary results of the simulations conducted on the energy damping components of the sabot-stopping system, some critical design issues have been identified that need to be addressed before the LGG can be fully implemented and used for testing and research purposes.

Author Contributions: Conceptualization, L.B.; Methodology, L.B., M.W. and T.S.; Formal analysis, L.B., M.W. and T.S.; Writing—original draft, L.B.; Writing—review & editing, L.B. and M.W. All authors have read and agreed to the published version of the manuscript.

Funding: This research received no external funding.

Institutional Review Board Statement: Not applicable.

Informed Consent Statement: Not applicable.

Data Availability Statement: Data is contained within the article.

Acknowledgments: This activity has been performed in the framework of a collaboration among the Department of Aviation at the Malta College of Arts, Science & Technology (MCAST), Malta; the Aerospace, Mechanical and Electronic Department at South East Technological University (SETU), Carlow campus, Ireland; and the Centre of Studies and Activities for Space (CISAS) "G. Colombo" of the University of Padova, Italy. The authors wish to thank Lorenzo Cocola, Roberto Tiscio, and Marco Chiaradia for the support to the design process.

Conflicts of Interest: The authors declare no conflict of interest.

Acronyms/Abbreviations

CISAS	Centre of Studies and Activities for Space
DOF	Degrees of Freedom
FEA	Finite Element Analysis
FEM	Finite Element Model
HVI	Hypervelocity impact
IADC	Inter-Agency Space Debris Committee
GEO	Geostationary Orbit
LEO	Low Earth Orbit
LGG	Light-Gas Gun
MCAST	Malta College of Arts, Science and Technology
MMOD	Micrometeoroids and Orbital Debris
SETU	South East Technological University

References

1. Kessler, D.J.; Cour-Palais, B.G. Collision frequency of artificial satellites: The creation of a debris belt. *J. Geophys. Res. Atmos.* **1978**, *83*, 2637–2646. [CrossRef]
2. Kessler, D.J.; Johnson, N.L.; Liou, J.-C.; Matney, M. The Kessler syndrome: Implications to future space operations. *Adv. Astronaut. Sci.* **2010**, *137*, 2010.
3. Anselmo, L.; Rossi, A.; Pardini, C. Updated results on the long-term evolution of the space debris environment. *Adv. Space Res.* **1999**, *23*, 201–211. [CrossRef]
4. Karacalioglu, A.G.; Stupl, J. *The Impact of New Trends in Satellite Launches on the Orbital Debris Environment*; NASA Technical Report; NASA: Houston, TX, USA, 2016. Available online: https://ntrs.nasa.gov/citations/20160011184 (accessed on 14 April 2023).
5. Virgili, B.B.; Dolado, J.; Lewis, H.; Radtke, J.; Krag, H.; Revelin, B.; Cazaux, C.; Colombo, C.; Crowther, R.; Metz, M. Risk to space sustainability from large constellations of satellites. *Acta Astronaut.* **2016**, *126*, 154–162. [CrossRef]
6. Olivieri, L.; Francesconi, A. Large constellations assessment and optimization in LEO space debris environment. *Adv. Space Res.* **2019**, *65*, 351–363. [CrossRef]
7. Lewis, H.; Radtke, J.; Rossi, A.; Beck, J.; Oswald, M.; Anderson, P.; Virgili, B.B.; Krag, H. Sensitivity of the space debris environment to large constellations and small satellites. *J. Br. Interplanet. Soc.* **2017**, *70*, 105–117.
8. Krag, H.; Serrano, M.; Braun, V.; Kuchynka, P.; Catania, M.; Siminski, J.; Schimmerohn, M.; Marc, X.; Kuijper, D.; Shurmer, I.; et al. A 1 cm space debris impact onto the Sentinel-1A solar array. *Acta Astronaut.* **2017**, *137*, 434–443. [CrossRef]
9. Kelso, T.S. Analysis of the Iridium 33 Cosmos 2251 Collision. *Adv. Astronaut. Sci.* **2009**, *135*, 1099–1112.
10. McKnight, D.; Di Pentino, F.; Knowles, S. Massive Collisions In LEO—A Catalyst to Initiate ADR. In Proceedings of the 65th International Astronautical Congress, Toronto, CA, USA, 29 September–3 October 2014.
11. Mark, C.P.; Kamath, S. Review of Active Space Debris Removal Methods. *Space Policy* **2019**, *47*, 194–206. [CrossRef]

12. Tadini, P.; Tancredi, U.; Grassi, M.; Anselmo, L.; Pardini, C.; Francesconi, A.; Branz, F.; Maggi, F.; Lavagna, M.; DeLuca, L.; et al. Active debris multi-removal mission concept based on hybrid propulsion. *Acta Astronaut.* **2014**, *103*, 26–35. [CrossRef]
13. Lorenzo, O.; Valmorbida, A.; Sarego, G.; Lungavia, E.; Vertuani, D.; Lorenzini, E.C. Test of tethered deorbiting of space debris. *Adv. Astronaut. Sci. Technol.* **2020**, *3*, 115–124.
14. Tarabini Castellani, L.; García González, S.; Ortega, A.; Madrid, S.; Lorenzini, E.C.; Olivieri, L.; Sarego, G.; Brunello, A.; Valmorbida, A.; Tajmar, M.; et al. Deorbit kit demonstration mission. *J. Space Saf. Eng.* **2022**, *9*, 165–173. [CrossRef]
15. Gleason, M.P. Establishing space traffic management standards, guidelines and best practices. *J. Space Saf. Eng.* **2020**, *7*, 426–431. [CrossRef]
16. Li, D. Upgrading space debris mitigation measures to cope with proliferating cyber threats to space activities. *Adv. Space Res.* **2023**, *71*, 4185–4195. [CrossRef]
17. Braun, V.; Braun, V.; Horstmann, A.; Lemmens, S.; Wiedemann, C.; Böttcher, L. Recent developments in space debris environment modelling, verification and validation with MASTER. In Proceedings of the 8th European Conference on Space Debris, Darmstadt, Germany, 20–23 April 2021; ESA Space Debris Office: Darmstadt, Germany, 2021.
18. Duzellier, S.; Gordo, P.; Melicio, R.; Valério, D.; Millinger, M.; Amorim, A. Space debris generation in GEO: Space materials testing and evaluation. *Acta Astronaut.* **2021**, *192*, 258–275. [CrossRef]
19. Olivieri, L.; Giacomuzzo, C.; Francesconi, A.; Stokes, H.; Rossi, A. Experimental characterization of multi-layer 3D-printed shields for microsatellites. *J. Space Saf. Eng.* **2020**, *7*, 125–136. [CrossRef]
20. Carrasquilla, R.E.; Norman, F.-C. DebriSat: Generating a dataset to improve space debris models from a laboratory hypervelocity experiment. *Europe* **2019**, *1*, 5–638.
21. Krisko, P.; Horstman, M.; Fudge, M. SOCIT4 collisional-breakup test data analysis: With shape and materials characterization. *Adv. Space Res.* **2008**, *41*, 1138–1146. [CrossRef]
22. Liou, J.C.; Fitz-Coy, N.; Clark, S.; Werremeyer, M.; Huynh, T.; Sorge, M.; Polk, M.; Roebuck, B.; Rushing, R.; Opiela, J. DebriSat–A planned laboratory-based satellite impact experiment for breakup fragment characterization. In Proceedings of the Sixth European Conference on Space Debris, Darmstadt, Germany, 22–25 April 2013.
23. Hanada, T.; Liou, J.-C. Comparison of fragments created by low- and hyper-velocity impacts. *Adv. Space Res.* **2008**, *41*, 1132–1137. [CrossRef]
24. Hanada, T.; Liou, J.-C.; Nakajima, T.; Stansbery, E. Outcome of recent satellite impact experiments. *Adv. Space Res.* **2009**, *44*, 558–567. [CrossRef]
25. Abdulhamid, H.; Bouat, D.; Coll'e, A.; Lafite, J.; Limido, J.; Midani, I.; Papy, J.-M.; Puillet, C.; Spel, M.; Unfer, T. On-ground HVI on a nanosatellite. Impact test, fragments recovery and characterization, impact simulations. In Proceedings of the 8th European Conference on Space Debris, Darmstadt, Germany, 20–23 April 2021.
26. Olivieri, L.; Smocovich, P.A.; Giacomuzzo, C.; Francesconi, A. Characterization of the fragments generated by a Picosatellite impact experiment. *Int. J. Impact Eng.* **2022**, *168*, 104313. [CrossRef]
27. Olivieri, L.; Giacomuzzo, C.; Francesconi, A. Analysis of fragments larger than 2 mm generated by a picosatellite fragmentation experiment. *Acta Astronaut.* **2023**, *204*, 418–424. [CrossRef]
28. Johnson, N.L.; Krisko, P.H.; Liou, J.-C.; Anz-Meador, P.D. NASA's new breakup model of EVOLVE 4.0. *Adv. Space Res.* **2001**, *28*, 1377–1384. [CrossRef]
29. McKnight, D.; Maher, R.; Nagl, L. Fragmentation Algorithms for Strategic and Theater Targets (FASTT) Empirical Breakup Model, Ver 3.0. DNA-TR-94-104. December 1994.
30. Sorge, M.E. Satellite fragmentation modeling with IMPACT. In Proceedings of the AIAA/AAS Astrodynamics Specialist Conference and Exhibit, Honolulu, HI, USA, 18–21 August 2008.
31. Sorge, M.E.; Mains, D.L. IMPACT fragmentation model developments. *Acta Astronaut.* **2016**, *126*, 40–46. [CrossRef]
32. Francesconi, A.; Giacomuzzo, C.; Olivieri, L.; Sarego, G.; Duzzi, M.; Feltrin, F.; Valmorbida, A.; Bunte, K.D.; Deshmukh, M.; Farahvashi, E.; et al. CST: A new semi-empirical tool for simulating spacecraft collisions in orbit. *Acta Astronaut.* **2019**, *160*, 195–205. [CrossRef]
33. Canning, T.N.; Seiff, A.; James, C.S. Ballistic Range Technology. In *AGARDograph*; Advisory Group for Aerospace Research and Development: Neuilly-Sur-Seine, France, 1970; Volume 138, p. 87.
34. United Nations Office for Outer Space Affair. Space Debris Mitigation Guidelines of the Committee on the Peaceful Uses of Outer Space. 2010. Available online: https://www.unoosa.org/pdf/publications/st_space_49E.pdf (accessed on 5 June 2023).
35. Bogdanoff, D.W. Optimization study of the Ames 0.5 two-stage light gas gun. *Int. J. Impact Eng.* **1997**, *20*, 131–142. [CrossRef]
36. Khristenko, Y.F. New light-gas guns for studying high-velocity impact at space velocities. In *Proceedings of the AIP Conference Proceedings*; AIP Publishing LLC.: Melville, NY, USA, 2017; Volume 1893. [CrossRef]
37. Angrilli, F.; Pavarin, D.; De Cecco, M.; Francesconi, A. Impact facility based upon high frequency two-stage light-gas gun. *Acta Astronaut.* **2003**, *53*, 185–189. [CrossRef]
38. Barilaro, L.; Olivieri, L.; Tiscio, R.; Francesconi, A. Evaluation of a Single-Stage Light-Gas Gun Facility in Malta: Business Analysis and Preliminary Design. *Aerotec. Missili Spaz.* **2022**, *101*, 159–169. [CrossRef]
39. Pavarin, D.; Francesconi, A. Improvement of the CISAS high-shot-frequency light-gas gun. *Int. J. Impact Eng.* **2003**, *29*, 549–562. [CrossRef]

40. Francesconi, A.; Pavarin, D.; Bettella, A.; Angrilli, F. A special design condition to increase the performance of two-stage light-gas guns. *Int. J. Impact Eng.* **2008**, *35*, 1510–1515. [CrossRef]
41. Oltrogge, D.; Alfano, S.; Law, C.; Cacioni, A.; Kelso, T. A comprehensive assessment of collision likelihood in Geosynchronous Earth Orbit. *Acta Astronaut.* **2018**, *147*, 316–345. [CrossRef]
42. Slimane, S.A.; Slimane, A.; Guelailia, A.; Boudjemai, A.; Kebdani, S.; Smahat, A.; Mouloud, D. Hypervelocity impact on honeycomb structure reinforced with bi-layer ceramic/aluminum facesheets used for spacecraft shielding. *Mech. Adv. Mater. Struct.* **2021**, *29*, 4487–4505. [CrossRef]
43. Plassard, F.; Mespoulet, J.; Hereil, P. Analysis of a single stage compressed gas launcher behaviour: From breech opening to sabot separation. In *Proceedings of the 8th European LS-DYNA Conference*; Livermore Software Technology Corporation (LSTC): Strasburg, France, 2011; Volume 8.
44. Tang, W.; Wang, Q.; Wei, B.; Li, J.; Li, J.; Shang, J.; Zhang, K.; Zhao, W. Performance and Modeling of a Two-Stage Light Gas Gun Driven by Gaseous Detonation. *Appl. Sci.* **2020**, *10*, 4383. [CrossRef]
45. Pavarin, D.; Lambert, M.; Francesconi, A.; Destefanis, R. Analysis of goce's disturbances induced by hypervelocity impact. In Proceedings of the SP-587 Fourth European Conference on Space Debris, Darmstadt, Germany, 18–20 April 2005.
46. Ioilev, A.G.; Bebenin, G.V.; Kalmykov, P.N.; Shlyapnikov, G.P.; Lapichev, N.V.; Salnikov, A.V.; Sokolov, S.S.; Motlokhov, V.N. A LGG arrangement for cut-off of the projectile sabot. In Proceedings of the SP-672 Fifth European Conference on Space Debris, Darmstadt, Germany, 12 February 2018.
47. Hibbert, R.; Cole, M.; Price, M.; Burchell, M. The Hypervelocity Impact Facility at the University of Kent: Recent Upgrades and Specialized Capabilities. *Procedia Eng.* **2017**, *204*, 208–214. [CrossRef]
48. Barilaro, L. Measurement Techniques for Assessing and Reducing the Risk Posed by Micrometeoroid and Orbital Debris to Space Vehicles. Ph.D. Thesis, University of Padova, Padova, Italy, 2012.
49. Jacquelin, E.; Pashah, S.; Lainé, J.; Massenzio, M. Estimation of the Impact Duration for Several Types of Structures. *Shock. Vib.* **2012**, *19*, 597–608. [CrossRef]

Disclaimer/Publisher's Note: The statements, opinions and data contained in all publications are solely those of the individual author(s) and contributor(s) and not of MDPI and/or the editor(s). MDPI and/or the editor(s) disclaim responsibility for any injury to people or property resulting from any ideas, methods, instructions or products referred to in the content.

Article

Numerical Modelling of High-Speed Loading of Periodic Interpenetrating Heterogeneous Media with Adapted Mesostructure

Alexander Kraus, Andrey Buzyurkin, Ivan Shabalin and Evgeny Kraus *

Khristianovich Institute of Theoretical and Applied Mechanics SB RAS, Novosibirsk 630090, Russia; akraus@itam.nsc.ru (A.K.); buzjura@itam.nsc.ru (A.B.); shabalin@itam.nsc.ru (I.S.)
* Correspondence: kraus@itam.nsc.ru

Abstract: A series of calculations has been conducted to study the high-speed interaction of space debris (SD) particles with screens of finite thickness. For the first time, taking into account the fracture effects, a numerical solution has been obtained for the problem of high-velocity interaction between SD particles and a volumetrically reinforced penetrating composite screen. The calculations were performed using the REACTOR 3D software package in a three-dimensional setup. To calibrate the material properties of homogeneous screens made of aluminum alloy A356, stainless steel 316L, and multilayer screens, methodical load calculations were carried out. The properties of materials have been verified based on experimental data through systematic calculations of the load on homogeneous screens made of aluminum alloy A356, stainless steel 316L, and multilayer screens comprising a combination of aluminum and steel plates. Several options for the numerical design of heterogeneous screens based on A356 and 316L were considered, including interpenetrating reinforcement with steel inclusions and a gradient distribution of steel throughout the thickness of an aluminum matrix. The study has revealed that the screens constructed as a two-layer composite of A356/316L, volumetrically reinforced composite screens, and heterogeneous screens with a direct gradient distribution of steel in the aluminum matrix provide protection for devices from both a single SD particle and streams of SD particles moving at speeds of up to 6 km/s. SD particles were modeled as spherical particles with a diameter of 1.9 mm made of the aluminum alloy Al2017-T4 with a mass of 10 mg.

Keywords: heterogeneous material; reinforcement; high-speed interaction; barrier of finite thickness; space debris

Citation: Kraus, A.; Buzyurkin, A.; Shabalin, I.; Kraus, E. Numerical Modelling of High-Speed Loading of Periodic Interpenetrating Heterogeneous Media with Adapted Mesostructure. *Appl. Sci.* 2023, *13*, 7187. https://doi.org/10.3390/app13127187

Academic Editor: Jérôme Morio

Received: 26 April 2023
Revised: 9 June 2023
Accepted: 13 June 2023
Published: 15 June 2023

Copyright: © 2023 by the authors. Licensee MDPI, Basel, Switzerland. This article is an open access article distributed under the terms and conditions of the Creative Commons Attribution (CC BY) license (https://creativecommons.org/licenses/by/4.0/).

1. Introduction

Man-caused pollution of near-Earth space poses a threat to spacecraft, including the real risk of damage and destruction. The term "space debris" (SD) appeared in the late 1980s and, according to Flury's definition [1], describes all artificial objects and their parts in near-Earth space that do not serve any useful purpose or function. Since the launch of the first artificial Earth satellite, the number of launches has increased, resulting in the accumulation of both large fragments and small SD particles [2].

The collision of spacecraft construction elements with SD may lead to catastrophic consequences or local damage resulting in a loss of efficiency or function. In addition, natural meteoric particles from distant space pose a threat. The longer the spacecraft is in flight, the greater the probability of a collision with SD particles. To ensure effective protection against various external influences, the spacecraft's body must be technologically advanced in production and have the smallest possible mass. For low-orbit spacecraft, designing hulls and protective screens is particularly relevant due to the concentration of a large number of SD on low orbits.

Several types of screens have been developed to protect against SD particles. Whipple's protection [3] is an innovative approach that prevents impacts by incorporating a large number of thin shells that destroy incoming kinetic energy particles. This creates a cloud of fragments that dissipate the kinetic energy of impact and distribute it over a larger area [4]. Multilayer heterogeneous screens are another type of protective construction that weakens the impact impulse through multiple reflections on multiple boundaries [5]. Other protective constructions, such as porous structures [6] and woven composites [7], have gaps in material properties that contribute to the dissipation of impact energy and prevent its spread. However, these protection technologies are often bulky, which presents a challenge in the design of aerospace systems where size and weight are severely limited. To sum up, the challenge of reducing the mass of spacecraft protective screens while maintaining their effectiveness remains a significant concern.

Since the dawn of the space era, composites based on organic matrices and metal matrices (MMC) have been developed for use in space. These materials possess high specific stiffness and a virtually zero coefficient of thermal expansion (CTE). Organic matrix composites, such as graphite/epoxy resin, have been used in the space for truss elements, paneling, antennas, waveguides, and parabolic reflectors. On the other hand, MMCs are able to withstand high temperatures and possess high thermal conductivity, low CTE, high specific stiffness, and strength. These potential advantages generated optimism regarding the use of MMCs in critical space systems in the late 1980s [8].

The successful development of many modern branches of technology is primarily associated with the utilization of cutting-edge materials in construction, certain parts of which are subjected to extreme loads due to various reasons. Making an appropriate choice of a material capable of enduring the applied stress over a specific period of time can be quite laborious without conducting a theoretical prediction. The stress–strain and thermodynamic calculations of the loaded material, in the hands of an experienced researcher, serve as the key to enhancing reliability and prolonging the lifespan of the entire structure.

The advancement and refinement of material creation technologies with predetermined properties, such as additive technologies, electron beam welding, and so on, have broadened the scope of heterogeneous materials' applications. Experimental work focused on practically developing manufacturing technologies for such heterogeneous mediums with specific properties far surpasses the methods used for predicting the properties themselves. Consequently, there now exists a significant gap between the practical implementation of complex heterogeneous materials and the level of knowledge concerning the properties of such materials under intense dynamic loads.

The development of additive manufacturing technologies for the production of structural elements stimulates new approaches to designing materials and products [9–13]. One of the current challenges at the intersection of mechanics, materials science, and physics is the development of methods and approaches for designing products with a certain material structure that provides the required functional and structural properties.

In mechanical engineering, there is a task to increase the strength and damping properties of metals, which often contradict each other. Additively manufactured interpenetrating composites are a new type of "metal-metal" composites for use in high-energy absorption systems. In this system, the matrix phase—a liquid metal with a melting temperature lower than the melting temperature of the lattice—is poured into the reinforcing phase with a continuous lattice configuration made additively. The result is a periodically interpenetrating composite in which each component forms a continuous network. Studies of such materials show that the boundary between the reinforcing and matrix phases can demonstrate significantly different mechanical properties of the composite, which allows for the dissipation of the impact energy.

For example, ref. [14] proposed a method for creating an Mg-NiTi composite with a bicontinuous architecture of an interpenetrating heterogeneous medium. For this, a magnesium melt is infiltrated into a three-dimensionally printed nitinol frame, which allows for the creation of a composite with a unique combination of mechanical properties: increased strength at various temperatures, remarkable damage resistance, good damping ability at various amplitudes, and exceptional energy absorption efficiency. After deformation, the shape and strength of the composite can be restored by thermal treatment.

Our scientific interest was sparked by an experiment studying the response of a heterogeneous medium with an adapted interpenetrating mesostructure. This mesostructure was fabricated using the hybrid additive technology known as PrintCast [15–17]. The experiment focused on investigating the impact loads that arise during high-velocity interactions, which can be utilized as one of the spacecraft protection options [18]. The results of the experiment demonstrated that the composite made of stainless steel 316L and aluminum alloy A356 using PrintCast technology is more resistant to delamination than monolithic screens made of the same materials. The studies have shown that the metal matrix composite with an adapted interpenetrating mesostructure is a promising system for spacecraft protection in cases where size limitations prevent the use of traditional protection methods. The heterogeneous mesostructure of this composite leads to a significant attenuation of the shock wave by multiple scattering at the interfaces of dissimilar materials and prevents macroscopic spall [19].

In work [20], the mechanical properties of PrintCast composites and their dependence on the volume fraction of reinforcement with steel 316L were studied. Uniaxial tensile tests were conducted on A356/316L PrintCast composites containing 30%, 40%, and 50% reinforcement. An increase in ductility by 200% and absorbed energy by 400% was observed when the volume fraction of reinforcement increased from 30% to 40%. However, with an increase in reinforcement from 40% to 50%, a much smaller increase in these parameters was observed. The sample with a volume fraction of 30% failed due to localized deformation in a single area after the onset of failure, unlike the samples with volume fractions of 40% and 50%, where the failure occurred due to non-localized damage throughout the cross-section of the sample.

The authors proposed the technology of direct numerical construction of heterogeneous media in [21,22], and comprehensive studies were conducted to determine the parameters in heterogeneous media, demonstrating their possible advantages. The works include comparisons with experimental data as well as descriptions of some mixture laws and methodologies for working with them. The work in [23] showed that during the propagation of an impulse in all media, it evolves into an elastic stress–strain state where its amplitude and length no longer depend on the distance traveled. Studies were also conducted on the influence of inclusion sizes. It was found that for heterogeneous materials with large inclusions, the rate of attenuation of the impulse amplitude is significantly higher compared to heterogeneous materials with small inclusions. Reducing the overall concentration of ceramics from 40% to 20% volume fraction in the heterogeneous material preserves all trends in the behavior of a short impulse during its propagation through an obstacle.

2. Mathematical Problem Statement

The software package "REACTOR 3D" [24] was used to perform the calculations. The Lagrangian approach is commonly employed to describe the dynamic interaction of deformable solids as it provides a suitable framework for characterizing the behavior of the medium. The region containing the composite is covered by a finite difference grid, where triangular-shaped cells fill the space without gaps or overlaps. Each triangular cell is assigned its own material's physical and mechanical properties. When transitioning from one cell to another, the characteristics can change abruptly. The boundaries of the cells satisfy conditions for the collective motion of the heterogeneous material components. Inside the cells, the investigated quantities are determined using an explicit finite difference scheme.

2.1. The Main Conservation Laws

The mathematical formulation is described in [22,25,26]. The partial differential equations are converted to an explicit difference scheme on the difference grid along the trajectory of each material particle. The procedure for constructing the difference scheme is described in detail elsewhere [25,26]. We use the system of equations for the deformable solid model, which includes the following equations:

- The particle trajectory equation

$$\dot{x}_i = u_i; \quad (1)$$

- The mass balance equation

$$V_0 \rho_0 = V \rho; \quad (2)$$

- The momentum balance equation

$$\rho \dot{u}_i = \sigma_{ij,j}; \quad (3)$$

- The internal energy balance equation

$$\rho \dot{e} = \sigma_{ij} \dot{\varepsilon}_{ij}; \quad (4)$$

$\dot{\varepsilon}_{ij}$ is the strain rate tensor:

$$\dot{\varepsilon}_{ij} = \frac{1}{2}(u_{i,j} + u_{j,i}); \quad (5)$$

σ_{ij} is the stress tensor:

$$\sigma_{ij} = -\delta_{ij} P + s_{ij}; \quad (6)$$

where s_{ij} is the stress deviator, which characterizes the shear-induced change in the shape of a material particle; δ_{ij} is the Kronecker symbol.

The elastoplastic flow equations are formulated in the form of Prandtl–Reuss equations.

$$\dot{s}_{ij} + d\lambda' s_{ij} = 2G \dot{\varepsilon}'_{ij}, \quad \dot{\varepsilon}'_{ij} = \dot{\varepsilon}_{ij} - \dot{\varepsilon}_{kk}/3, \quad (7)$$

with the Huber–von Mises plasticity condition

$$s_{ij} \cdot s_{ij} \leq 2 \cdot Y_0^2 / 3, \quad (8)$$

where Y_0 is the dynamic yield stress. Instead of calculating the scalar factor $d\lambda'$, we use the well-known procedure of reducing the stress deviator components to the yield circle [26]. In Equations (1)–(8), each of the subscripts i, j takes values 1, 2, and 3; summation is performed over repeating indices; a dot above a symbol denotes the time derivative, and a subscript after a comma denotes the derivative with respect to the corresponding coordinate; x_i and u_i are the components of the position and velocity vectors of a material particle, respectively; ρ is the current density; G is the shear modulus.

2.2. The Equation of State

A few-parameter equation of state in the form of the Mie–Gruneisen equation [27,28] is used

$$P = P_x + \frac{\gamma(V) c_{v,l} T}{V} + \frac{c_{v,e} T^2}{3V(V/V_0)^{2/3}}.$$

Here, P_x is the pressure on the zero isotherm; T is the temperature; $c_v = c_{v,l} + c_{v,e}$ is the constant-volume heat capacity equal to the sum of the lattice and electronic heat capacities; V and V_0 are the current and initial specific volumes; $\gamma(V)$ is the Gruneisen coefficient.

2.3. The Boundary Conditions

Each body in the Cartesian coordinate system $\{x_j\}$ corresponds to a computational domain $D^i(\mathbf{x}, t)$ with boundaries $G^i(\mathbf{y}, t)$ (see Figure 1). Here, $\mathbf{x} = \mathbf{x}(t)$ is the position vector of the material particle and **y** are the boundary points. In the general case, the computational domains $D^i(\mathbf{x}, t)$ change in time and can be multi-connected.

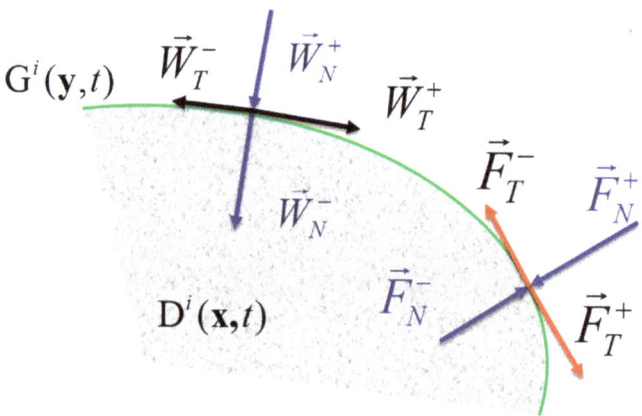

Figure 1. Boundary conditions.

The initial conditions for the i-th body at $t = 0$ in the region $D^i(\mathbf{x}, t)$ are of the form:

$$\rho^i(\mathbf{x}, 0) = \rho^{i0}(\mathbf{x}), \ u^i_j(\mathbf{x}, 0) = u^{i0}_j(\mathbf{x}), \ s_{ij} = P = e = 0,$$

where $\rho^{i0}(\mathbf{x})$ and $u^{i0}_j(\mathbf{x})$ are the given initial distributions of the material density and the velocity vector over the area $D^i(\mathbf{x}, t)$.

To formulate the boundary conditions, let us introduce the following notations:

- n^i the vector of the outward normal to the boundary $G^i(\mathbf{y}, t)$ of the domain $D^i(\mathbf{x}, t)$;
- $F^i(\mathbf{y}, t)$ the vector of external surface forces on the boundary $G^i(\mathbf{y}, t)$;
- $W^i(\mathbf{y}, t)$ the vector of velocity at the boundary. $G^i(\mathbf{y}, t)$.

The following conditions can be specified on any boundary of the domain $G^i(\mathbf{y}, t)$:

- kinematic

$$\mathbf{u}^i(\mathbf{y}, t) = W^i(\mathbf{y}, t),$$

- dynamic

$$\sigma^i_{kl}(\mathbf{y}, t) \, n^i_l = F^i_k(\mathbf{y}, t), \ (k, l = 1, 2, 3),$$

where σ^i_{kl} are the components of the stress tensor on the boundary of the domain $G^i(\mathbf{y}, t)$, which typically needs to be determined;

- mixed

$$\mathbf{u}^i(\mathbf{y}, t) = W^i(\mathbf{y}, t), \ \mathbf{y} \in G^{i\alpha}(\mathbf{y}, t), \ \sigma^i_{kl}(\mathbf{y}, t) n^i_l = F^i_k(\mathbf{y}, t), \mathbf{y} \in G^{i\beta}(\mathbf{y}, t), \ G^i(\mathbf{y}, t) = G^{i\alpha}(\mathbf{y}, t) \cup G^{i\beta}(\mathbf{y}, t).$$

The contact surface between two bodies $G^{ij}(\mathbf{z}, t)$ is defined as the set of points $\mathbf{z}$ that satisfy the condition.

$$G^{ij}(\mathbf{z}, t) = G^i(\mathbf{y}, t) \cap G^j(\mathbf{y}, t).$$

Certain compatibility conditions must be satisfied on the contact boundary between the bodies for the vectors of velocities $\mathbf{u}^i(\mathbf{z}, t)$ and $\mathbf{u}^j(\mathbf{z}, t)$, as well as the components of the stress tensor σ^i_{kl} and σ^j_{kl}. Specific types of conditions on the contact boundaries will be stated below. To simplify the algorithm for calculating the motion of the boundaries, we will use the external surface force vectors $F^i(\mathbf{y}, t)$. The reaction forces, which are determined during the problem-solving process, will be denoted by $\mathbf{R}^i(\mathbf{z}, t)$.

Let us assume that at each point **z** on the contact surface $G^{ij}(\mathbf{z},t)$, there exists a common normal. In this case,

$$\mathbf{n}^i = -\mathbf{n}^j.$$

Let us decompose the vector $\mathbf{A}(\mathbf{z},t)$ at point **z** into its normal $\mathbf{A}_n$ and tangential $\mathbf{A}_t$ components

$$\mathbf{A} = \mathbf{A}_n + \mathbf{A}_t. \tag{9}$$

These components can be calculated using the formulas.

$$\mathbf{A}_n = (\mathbf{A}, \mathbf{n}) \cdot \mathbf{n}, \quad \mathbf{A}_t = \mathbf{n} \times (\mathbf{n} \times \mathbf{A}).$$

Let us replace the action of body i on body j at point **z** with the reaction force vector $\mathbf{R}^i(\mathbf{z},t)$, and, correspondingly, the action of body j on body i with the reaction force vector $\mathbf{R}^j(\mathbf{z},t)$. Then, $\mathbf{R}^i(\mathbf{z},t) = -\mathbf{R}^j(\mathbf{z},t)$, and according to (9), we will have

$$\mathbf{R}^i = \mathbf{N}^i + \mathbf{T}^i,$$

where $\mathbf{N}^i$ and $\mathbf{T}^i$ are the normal and tangential components of the reaction force vector, respectively.

Let us consider the formulation of boundary conditions on the contact surface in specific cases:

- Ideal mechanical contact: The material particles belonging to the boundaries of the interacting bodies move as a single entity.

$$\mathbf{u}^i(\mathbf{z},t) = \mathbf{u}^j(\mathbf{z},t), \quad \mathbf{R}^i(\mathbf{z},t) = -\mathbf{R}^j(\mathbf{z},t); \tag{10}$$

- Frictionless sliding: In this case, the conditions of non-penetration and equilibrium hold for the normal components of the reaction forces.

$$\mathbf{u}_n^i(\mathbf{z},t) = \mathbf{u}_n^j(\mathbf{z},t), \quad \mathbf{N}_n^i(\mathbf{z},t) = -\mathbf{N}_n^j(\mathbf{z},t), \quad \mathbf{T}^i(\mathbf{z},t) = \mathbf{T}^j(\mathbf{z},t) = 0, \quad \mathbf{z} \in G^{ij}(\mathbf{z},t); \tag{11}$$

Condition (11) is applied only for compressive reaction forces, i.e., $(\mathbf{N}^i, \mathbf{n}^i) < 0$. If this condition is violated, stress-free surface conditions are applied to the boundaries $G^i(\mathbf{y},t)$ and $G^j(\mathbf{y},t)$.

- Sliding with Coulomb friction: Let the friction coefficient be k. The friction force is determined by the expression,

$$\mathbf{T}^i = \kappa \left|\mathbf{N}^i\right| \mathbf{q}^i, \quad \mathbf{q}^i = -\frac{\mathbf{u}_t^i - \mathbf{u}_t^j}{\left|\mathbf{u}_t^i - \mathbf{u}_t^j\right|}, \text{ if } (\mathbf{N}^i, \mathbf{n}^i) < 0,$$

where $\mathbf{q}^i$ is the unit vector in the tangential plane to the contact surface, directed against the relative velocity vector. The boundary conditions take the form

$$\mathbf{u}_n^i(\mathbf{z},t) = \mathbf{u}_n^j(\mathbf{z},t), \quad \mathbf{R}^i(\mathbf{z},t) = -\mathbf{R}^j(\mathbf{z},t). \tag{12}$$

The tangential components of velocities are calculated based on the tangential components of the reaction force $\mathbf{T}^i(\mathbf{z},t)$, and their preliminary values $\mathbf{T}^{i*}(\mathbf{z},t)$ are obtained from the second Relation (10). If the magnitude of $|\mathbf{T}^{i*}(\mathbf{z},t)| < k\left|\mathbf{N}^i(\mathbf{z},t)\right|$, then $\mathbf{T}^i(\mathbf{z},t) = k\left|\mathbf{N}^i(\mathbf{z},t)\right|$. Otherwise, the internal forces cannot overcome the frictional forces, and there is no sliding at the interface. In that case, the sliding Condition (12) is replaced by the ideal contact Condition (10).

2.4. Fracture

Kinematic strength characteristics include the limiting values of elongation (usually under uniaxial tension) and shear. Brittle materials are also destroyed by compressive strains. Kinematic characteristics are accumulated quantities incorporating the entire history of the process. The most frequently used quantities are the limiting elongation and shear strain. Finding these quantities involves the calculation of the primary tensile and compressive strains.

$$\varepsilon_1 = \frac{\varepsilon_{xx}+\varepsilon_{yy}}{2} + \sqrt{\left(\frac{\varepsilon_{xx}-\varepsilon_{yy}}{2}\right)^2 + \varepsilon_{xy}^2},$$

$$\varepsilon_2 = \frac{\varepsilon_{xx}+\varepsilon_{yy}}{2} - \sqrt{\left(\frac{\varepsilon_{xx}-\varepsilon_{yy}}{2}\right)^2 + \varepsilon_{xy}^2}$$

and also, the shear strain

$$\varepsilon_\tau = \frac{(\varepsilon_1 - \varepsilon_2)}{2}.$$

If the tensile strains in the course of deformation exceed the limiting elongation ε_1^* (i.e., $\varepsilon_1 > \varepsilon_1^*$) or the limiting shear strain $\varepsilon_\tau > \varepsilon_\tau^*$, ε_τ^*, then the element material is assumed to be fractured, i.e., it has no longer resistance to tension and shear but still has resistance to compression.

Force strength parameters include the limiting values of tensile [29], compressive, and shear stresses. If the stresses proper are used, then they are instantaneous criteria, i.e., as soon as the principal stresses exceed the limiting values, the element material is assumed to be fractured:

$$\begin{cases} \sigma_1 > \sigma_1^*, \\ \sigma_2 > \sigma_2^*, \\ \sigma_\tau = \frac{(\sigma_1-\sigma_2)}{2} > \sigma_\tau^*. \end{cases}$$

Most materials, however, possess properties of plasticity and viscosity; therefore, their fracture requires a time interval during which the material is under overstrain. The code involves one of such criteria (it is demonstrated by an example of the principal tensile stress) [30]:

$$\sigma_t = \frac{\sum\limits_{i=n_1}^{n_2}(\sigma_1 - \sigma_1^*)_i \Delta t_i}{\sum\limits_{i=n_1}^{n_2}\Delta t_i} > \sigma_t^*, \quad \sigma_1 - \sigma_1^* > 0.$$

2.5. Conversion of Fractured Elements to Particles

If the element is at the boundary of the computational domain and the parameters of the material reach a critical value, the material of the element is replaced by discrete particles. The radius of the particle is calculated from the condition of incorporating one or more of the particles in the element. The mass of the element is allocated between the discrete particles. For a one-time step, only one layer of boundary elements can be converted into discrete particles, so the velocity of the wavefront of destruction does not exceed the speed of disturbances in the medium. Figure 2 shows the conversion of triangle elements (A, B, and others) to particles [31] in the 2D formulation:

- Element A is removed from the element grid;
- Particle A is added as a particle node;
- All of the element variables (stress, strain, damage, etc.) are transferred to the particle;
- The mass, velocity, and center of gravity of the particle node are set to those of the replaced element. The nodal velocity is obtained from the momentum of the element (three nodal masses and velocities);
- The masses of nodes b, c, and k are reduced by the removal of element A;
- For the conversion of element B (which has two sides on the surface) to node B, most of the steps are similar to those used for element A.

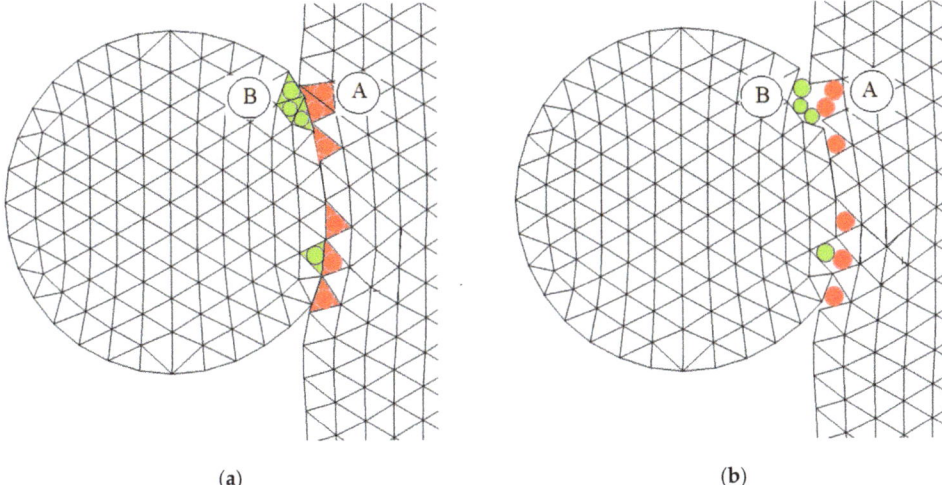

Figure 2. Conversion of triangle elements to particles. (**a**)—interface nodes and elements before conversion, (**b**)—interface after conversion of elements. The green circles and areas belong to Interacting Body 1, and the red ones to Interacting Body 2.

In the 3D formulation, the algorithm for replacing cells containing damaged material when they reach critical values with discrete particles that simulate the behavior of the destroyed material is identical to the 2D case described above.

The behavior of the shear modulus under pressure and temperature is described in [32]. The model of a heterogeneous environment and the solved problems are presented in [33–35].

3. Propagation of Shock Waves in a Periodically Volume-Reinforced Metal Matrix Composite

The problem of determining effective modules for heterogeneous media dates back to classical works [36,37], while research on shock waves has been conducted in [38,39]. Although this problem has been addressed in many monographs, such as [40,41], it has yet to find a final solution. In [42], a deviation from the rule of mixtures in the shock adiabat was discovered for a volume-reinforced metal matrix composite. As this phenomenon was not observed in our work, for example, in [23,33], we conducted numerical calculations of a volume-reinforced metal matrix composite using averaging according to the methodology outlined in [33]. Figure 3 shows a model of such a composite that we used in our calculations, similar to [18].

Figure 4 displays the calculated dependence of the shock wave velocity on the mass velocity, revealing that the direct numerical modeling of a penetrating heterogeneous medium demonstrates that the shock wave velocity corresponds to the calculation of the shock adiabat of the composite using the additive approach (rule of mixtures), at least at shock wave velocities exceeding C_l, the longitudinal sound speed in the composite, showing deviations from the parameters calculated using the mixture rule of no more than 3%.

Although numerous numerical studies of shock waves were carried out in an interpenetrating composite, we do not present them here as a complete analysis of the shock wave propagation in a periodically volume-reinforced metal matrix composite has already been conducted in [43]; the results of which fully consistent with our calculations. Only one question remains unanswered: the consideration of a reinforced composite fracture under high-speed loading and the influence of the heterogeneous screen structure on this process.

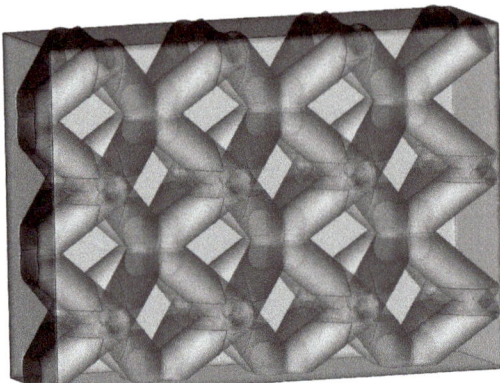

Figure 3. A model of a volume-reinforced composite with an A356 matrix and 316L inclusions.

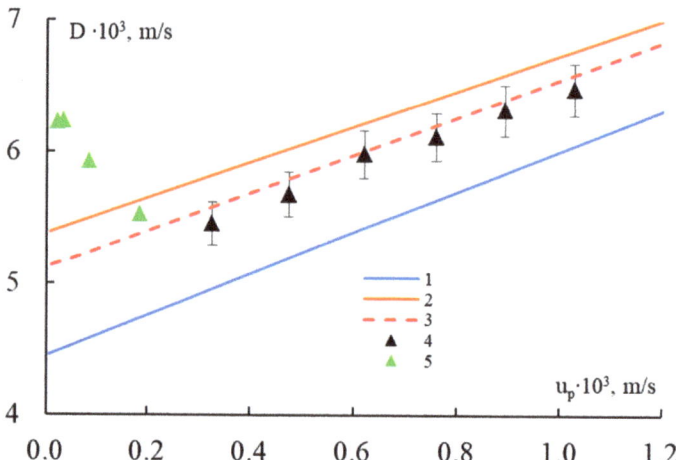

Figure 4. Shock wave velocity vs. the mass velocity: 1—316L steel; 2—A356 aluminum alloy; 3—calculation according to the additive rule of mixtures; 4—direct numerical modeling in a metal matrix composite, 5—C_l.

4. Problem Statement

Our objective is to design a heterogeneous screen to protect critical spacecraft components from space debris particles. We aim to investigate the high-speed interaction between spherical space debris particles and protective screens made of various structures using specified materials. The configuration of the problem is presented in Figure 5, which depicts the reinforced A356 + 316L screen placed in a protective casing to prevent its movement.

To calibrate the parameters of the computational model, we will utilize the work of [18], which provides experimental data on high-speed collisions of aluminum particles with five distinct screens. The numerical simulation of the collision processes permits to identification of the screens that exhibit adequate resilience against space debris particles. Additionally, we consider another three screens that can be produced using selective laser sintering of powders from the same materials as in [18]. The material properties for the aluminum alloys Al2017-T4 and A356 are taken from [44,45] and for steel 316L from [46].

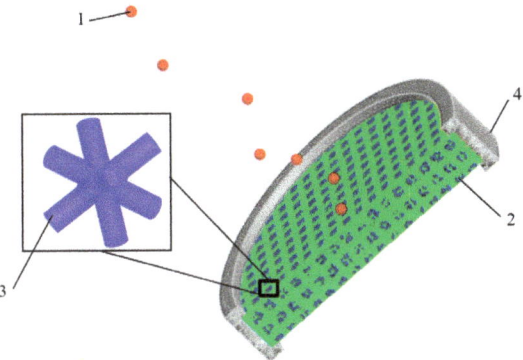

Figure 5. Geometric model: 1—Al2017-T4 spheres; 2—A356/316L volume-reinforced screen; 3—shape of the volume reinforcement; 4—heavy alloy protective casing.

For calculations, we will use the geometric dimensions of the impacting particle and screens from [18]. A spherical particle with a diameter of 1.9 mm made of the aluminum alloy Al2017-T4 with a mass of 10 mg moves at an initial speed of 6.0 km/s in all calculations. Every protective screen, with the exception of screen #2, has the same areal density. We evaluate the following protective screens:

- Screen #1—a steel 316L plate, 4.5 mm thick;
- Screen #2—an A356 aluminum alloy plate, 12.0 mm thick;
- Screen #3—a two-layer plate made of 316L/A356, 7.5 mm thick (3.0/4.5);
- Screen #4—a two-layer plate made of A356/316L, 7.5 mm thick (4.5/3.0);
- Screen #5—a metal-matrix composite plate with the matrix composed of the A356 aluminum alloy and the volume reinforcement made of steel 316L, as is illustrated in Figure 3. The unit volume of the heterogeneous inclusion in such a plate has a face-centered cubic (FCC) symmetry with a side length of 2.5 mm and a diameter of 0.8 mm. The screen thickness is 7.5 mm;
- Screens #6, #7, and #8 are considered later as alternatives;
- Screen #9—A356 aluminum alloy plate, 13.4 mm thick.

The diameter of the protective screens in all calculations is 50.0 mm. Every protective screen, with the exception of screen #2, has the same areal density. All screens are placed in a protective casing to eliminate the influence of free lateral surfaces on the stress state of the screen materials. The effective volumetric concentration of steel in screens #3, #4, and #5 is 38%. The thicknesses and areal densities of the screens are listed in Table 1.

Table 1. Parameters of the considered screens.

Screen	Material of the Screen	Density, g/cm^3	Areal Density, g/cm^2	Thickness, mm
#1 [18]	Stainless steel 316L	8.0	3.6	4.5
#2 [18]	Aluminum alloy A356	2.7	3.24	12.0
#3 [18]	Two-layer composite 316L/A356	4.8	3.6	7.5
#4 [18]	Two-layer composite A356/316L	4.8	3.6	7.5
#5 [18]	Volume-reinforced composite	4.8	3.6	7.5
#6	Uniform distribution of 316L steel grains in an A356 matrix	4.8	3.6	7.5
#7	Direct gradient distribution of 316L steel grains in an A356 matrix	4.8	3.6	7.5
#8	Inverse gradient distribution of 316L steel grains in an A356 matrix	4.8	3.6	7.5
#9	Aluminum alloy A356	2.7	3.6	13.4

5. Material Parameter Calibration

The numerical calculations gave the parameters for the craters (depth, diameter, and volume), which were compared with experimental data for all screens. As the calculated craters had a shape close to an ellipsoid, as seen in Figure 6a, an ellipse with semi-axes a and b was inscribed in the transverse section of the calculated crater to determine its volume by the formula:

$$V_{calc} = \pi \left(\frac{b}{a}\right)^2 h^2 \left(a - \frac{h}{3}\right)$$

In most calculations, an ellipsoid with semi-axes $a = h$ and $b = d/2$ can be inscribed in the crater, and the volume of the crater is calculated using the formula:

$$V_{calc} = \pi \, h \, d^2 / 6$$

The results of numerical modeling of the interaction of a spherical particle with screens of different structures were summarized, and the obtained crater parameters were compared with the experimental data from [18]. The experimental data and the results of numerical simulations of the formed craters for all considered screens are presented in Table 2, where h, d, and V are the depth, diameter, and volume of the crater as measured in the experiment, and h_{calc}, d_{calc}, and V_{calc} are the depth, diameter, and volume of the crater as obtained through numerical simulation.

Table 2. Crater size in the screens.

Screen Type	Experiment				Numerical Simulation			
	h, mm	d, mm	V, mm^3	Spall	h_{calc}, mm	d_{calc}, mm	V_{calc}, mm^3	Spall
#1	2.1	5.7	36	spall	2.30	5.60	31	spall
#2	4.5	8.2	158	spall	4.86	7.20	133	spall
#3	2.4	6.4	52	spall	3.70	5.35	55	spall
#4	4.5	8.4	166	no	4.25	7.40	162.6	no
#5	4.7	8.0	55	delamination	3.28	6.38	63	single grain detachment
#6	-	-	-	-	3.20	5.84	54	single grain detachment
#7	-	-	-	-	3.62	6.63	78	no
#8	-	-	-	-	3.01	5.47	39	spall
#9	-	-	-	-	4.67	7.42	134	spall

To determine the parameters of the calculated crater, a plate with a thickness of approximately the step of the division of the computational grid was cut out of the screen, passing through the center of the crater. The comparison between the calculated shape of the crater in screen #1, made of 316L steel, and the shape of the experimental crater is presented in Figure 6. Since the "REACTOR 3D" software package uses a hybrid meshless-mesh calculation method, the destroyed material is represented by finite-size particles, differentiated by color in all figures. It is worth noting that the depth of the calculated crater exceeds the depth of the experimental one by approximately 10%, while the calculated diameter almost coincides with the experimental one. However, there is a deviation in the volume of the crater; it reaches 14%. The thickness of the spalled part on the screen's backside is comparable to the experimental thickness.

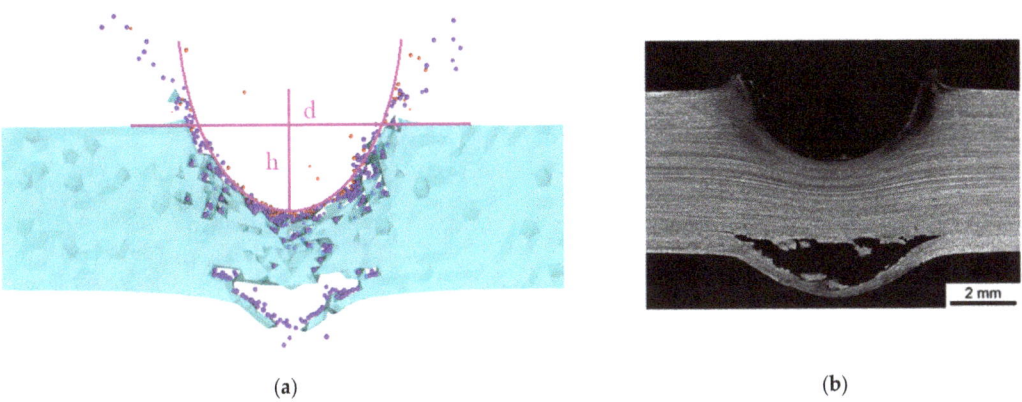

Figure 6. Comparison of the crater shapes in the 316L steel screen #1: (**a**) simulation results; (**b**) optical micrograph of the impact plane in cross-section [18].

The calculated crater in screen #2, made of aluminum alloy A356, was found to be 8% deeper, 12% smaller in diameter, and 16% smaller in volume than the experimental one, as shown in Figure 7. There are cracks on the front side of the screen which allow the "sponges" of the crater to separate from the screen. The thickness of the spall on the screen's backside is slightly smaller than the experimental one. Radial cracks caused by bending deformations are also present, as well as in the experiment.

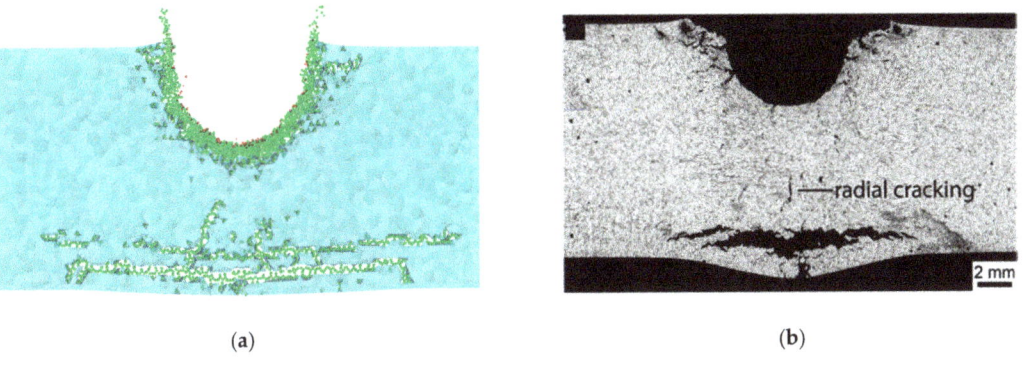

Figure 7. Comparison of the crater shapes in screen #2 of the aluminum alloy A356: (**a**) simulation results; (**b**) optical micrograph of the impact plane in cross-section [18].

To ensure consistency in areal density, screen #2, made from A356 aluminum alloy, should possess a thickness of 13.4 mm. We conducted calculations for screen #9 with the required thickness. The calculated crater parameters for screen #9 have values quite close to the experimental values (see Table 2). By increasing the thickness of screen #9 by 1.4 mm compared to screen #2, the depth of the crater decreased while its diameter increased. On a semi-infinite screen, the crater's shape would become hemispherical.

Let us examine a two-layer screen composed of aluminum alloy plates on the front and stainless steel on the back (screen #3). The stainless steel 316L plate is 3 mm thick, while the aluminum alloy A356 plate is 4.5 mm thick. In this case, the calculated crater parameters have several high errors: the depth exceeds the experimental depth by 12.5%, the diameter is smaller by 16.4%, and the volume of the crater exceeds the experimental volume by only 5.8%. Figure 8 shows a comparison of the crater cross-sections. In the calculation, the steel plate sustained significantly more damage compared to the experiment, while the aluminum plate suffered damage and fracture similar to the experimental sample. The only difference is that the plate has a larger diameter but slightly less thickness. It can be concluded that this is the upper limit of the process parameters for material fracture of the screen.

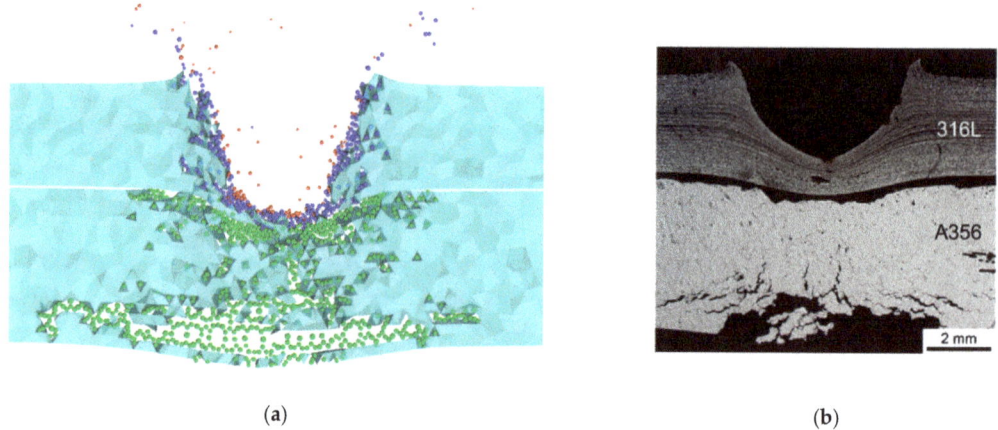

(a) (b)

Figure 8. Comparison of the crater shapes in screen #3—two-layer protection 316L + A356: (a) Simulation results; (b) Optical micrograph of the impact plane in cross-section [18].

For screen #4, made up of the A356 alloy plates and 316L steel, the calculated crater depth is 5.6% less than the experimental one, and the diameter is 12% smaller, but the volume of the crater is almost the same as the experimental one, with only a 2.1% deviation. There are no chips in the steel, nor are there any separate particles torn from the back surface. The calculated crater's distinctive feature is its almost cylindrical shape, which is due to the fragile aluminum alloy being crushed on the steel plate, as shown in Figure 9.

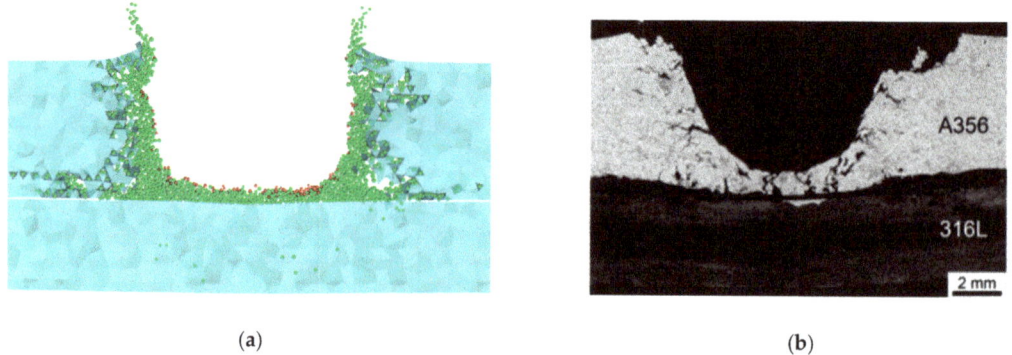

(a) (b)

Figure 9. Comparison of the crater shapes in screen #4—two-layer protection A356 + 316L: (a) simulation results; (b) optical micrograph of the impact plane in cross-section [18].

Thus, methodological calibration calculations were carried out using the physical and mechanical properties from [44–46] and experimental data from [18], giving the fracture parameters of homogeneous materials summarized in Table 3.

Table 3. Material parameters.

	K, GPa	G, GPa	C_0, m/s	S	Y, GPa	σ, GPa	ε, %
A356	73.95	26.00	5392	0.270	0.20	0.27	0.045
316L	130.00	79.00	4464	1.544	0.75	2.50	0.05
Al2017-T4	75.71	27.86	5538	1.338	0.28	0.456	0.12

6. Metal Matrix Protective Screens with Interpenetrating Periodic Inclusions

Based on the data obtained, we model a metal matrix reinforced screen with interpenetrating periodic inclusions with an adaptive mesostructure. Numerically constructing such a reinforced screen with volumetric interpenetrating periodic inclusions is a rather complex task. However, it is worth noting that the software package 'REACTOR 3D' allows for the construction of inclusions with an arbitrary 3D shape. In this case, the steel inclusions had the shape of a volumetric cross of four cylinders, as Figures 3 and 5 show. The comparison of the calculated crater's shape to that of the experimental crater in screen #5 is shown in Figure 10. The distinguishing feature of this screen is the stratification between the steel reinforcement and the aluminum matrix. Although there is no macroscopic spall on the backside of the screen, individual grains have been ejected due to spallation processes (refer to Figure 10a). These grains have a very small mass of approximately 0.1 mg and a velocity of around 50–70 m/s, making them an insignificant threat to protected devices.

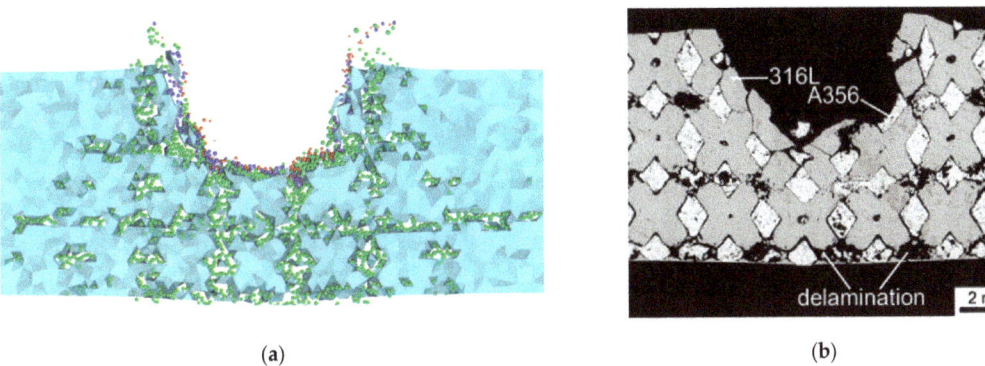

Figure 10. Craters in reinforced screen #5: (**a**) 3D calculation; (**b**) optical microphotograph of the impact plane in cross-section [18].

The calculated shape of the craters in the composite reinforced screen is close to the shape of the craters obtained experimentally. It is worth noting that there are significant discrepancies in the parameters, such as the crater depth, which is 30.2% less than the experimental value, and its diameter, which is 20.2% less than the experimental value. Even with the calculated values of the depth, which is equal to 3.28 mm, and the diameter, which is 6.38 mm, we obtain that the volume of the crater is 14.5% greater than the experimental volume. Apparently, there was a mistake in the parameters of the crater [18].

To quantitatively compare the calculated and experimental crater parameters, the results are tabulated in Table 2, while the deviation in the crater parameters is expressed as a percentage and tabulated in Table 4.

Table 4. Deviation of calculated crater parameters from experimental data.

Error/Screen	#1	#2	#3	#4	#5
Error depth %	9.5	8.0	12.5	5.6	30.2
Error diameter %	1.8	12.2	16.4	11.9	20.2
Error volume %	13.9	15.8	5.8	2.1	14.5

Besides the armor in the form of a volumetric cross of four cylinders, screens reinforced with periodic volumetric inclusions, as shown in Figure 11, were also considered. Discrete steel inclusions in the aluminum matrix in the form of cylinders and spheres (see Figure 11a,b), while maintaining the concentration of components, do not provide adequate protection, as they allow for penetration of the heterogeneous screen, and macroscopic spallation is observed on the backside. Armor screens consisting of "half-crosses" layers (see Figure 11c,d) and lattices (see Figure 11e), while maintaining the concentrations, do not allow for through penetration, but on the back side of such screens, spallation phenomena are observed in the form of the detachment of particles of the A356 aluminum alloy, which have a sufficiently high speed, about 400 m/s, posing a danger to protected devices.

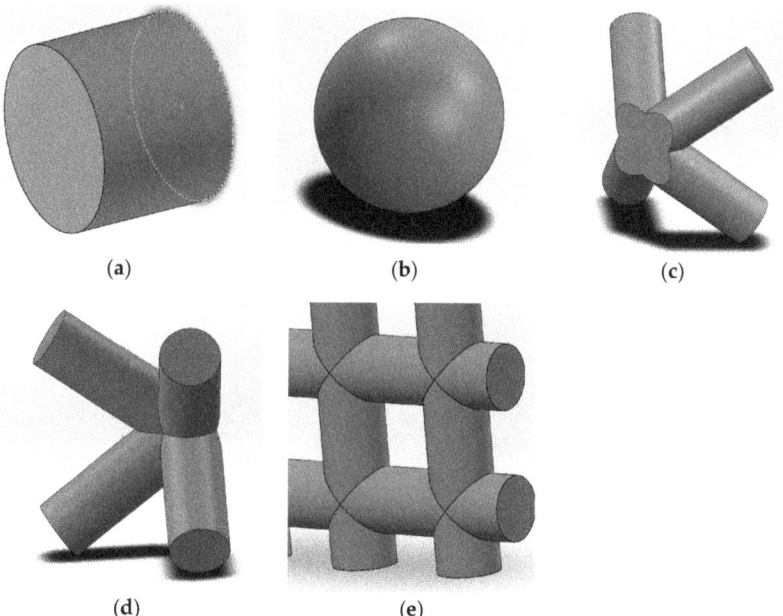

Figure 11. Discrete steel inclusions in the aluminum matrix: (**a**) cylinder; (**b**) sphere; (**c**) straight "half-cross"; (**d**) reverse "half-cross"; (**e**) lattice.

7. Gradient Protective Screens

To provide a complete overview of protective screen configurations, we additionally consider the following heterogeneous screens based on the A356 alloy and 316L steel with a volume content of 38% that can be created using existing additive technologies: (1) the screen #6 with a uniform distribution of steel throughout the screen volume (see Figure 12a); (2) the screen #7 with a direct gradient distribution of steel through the screen volume (see Figure 12b); and (3) the screen #8 with a reverse gradient distribution of steel through the screen volume (see Figure 12c). The heterogeneous screens with the gradient steel distribution are similar to layered barriers but with a continuous transition from one material to another without clearly defined boundaries.

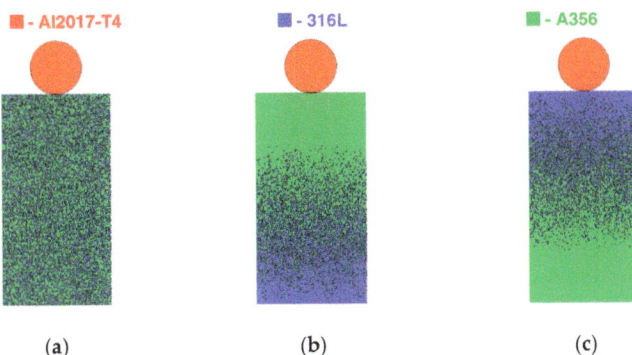

Figure 12. Cross-sections of heterogeneous screens with a steel concentration of 38%: (**a**) Screen #6 with a uniform distribution of 316L steel within the A356 aluminum alloy; (**b**) Screen #7 with a direct gradient distribution of steel; (**c**) Screen #8 with a reverse gradient distribution of steel.

Figure 13 presents the results of the impact loading calculations for spherical particles on the heterogeneous screens. Note that heterogeneous screens behave almost identically to two-layer ones. The uniform distribution of 316L steel in the aluminum matrix increases the effective yield strength, which has a positive effect on the crater volume reduction. However, the presence of aluminum matrix grains on the screen's reverse side leads to the formation of a stream of small particles moving at a speed of 150–250 m/s.

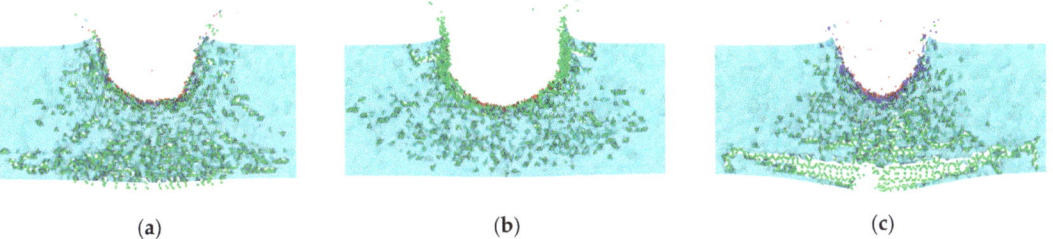

Figure 13. Calculation results of the interaction between an aluminum sphere and heterogeneous screens. Cross-sections of screens with a thickness equal to the grid step: (**a**) screen #6; (**b**) screen #7; (**c**) screen #8.

Screen #7, which has 100% A356 aluminum alloy on the front side and 100% 316L steel on the back side, has better protective properties than layered screen #4. This is reflected in the calculation of the crater parameters, such as the depth, which is 20% less, the diameter, which is 21% less, and the volume, which is 53% less. As with screen #4, there is no flow of microparticles on the back side, and there is no macroscopic spall.

The behavior of screen #8 closely resembles that of screen #3, as it displays a macroscopic spall with a detached fragment of considerable diameter. The screen volume has suffered significant material damage, particularly near the rear surface.

8. Multiple Impacts of Space Debris Particles

Since the service life of spacecraft is assumed to be 10–20 years, it is natural that protective screens should withstand multiple impacts from SD particles. Let us determine the mass of the incoming SD particles that the armored screen #5 can protect, avoiding penetration. To reduce the number of calculations, it is necessary to predict the maximum ballistic velocity of the screen from the mass of the incoming particle. Taylor's work [47] proposes a formula for the engineering estimation of the ballistic velocity:

$$\frac{1}{2}m_p V_{bl}^2 = \pi r^2 h Y, \qquad (13)$$

where m_p is the particle mass, V_{bl} is the ballistic velocity, r is the hole radius, h is the screen thickness, Y is the yield strength of the screen material.

As the ballistic velocity needs to be evaluated, let us assume that the combination of screen parameters is some constant value. Then, the expression for ballistic velocity is as follows:

$$V_{bl} = \frac{C}{\sqrt{m_p}}, \qquad (14)$$

To calibrate the constant C, two or three calculations need to be carried out to determine the ballistic velocity at a given particle mass. In our case, $C \sim 1.01$. Figure 14 illustrates the applicability of this formula and the results of numerical calculations. This Approximation (14) provides a 10% error margin for the calculation.

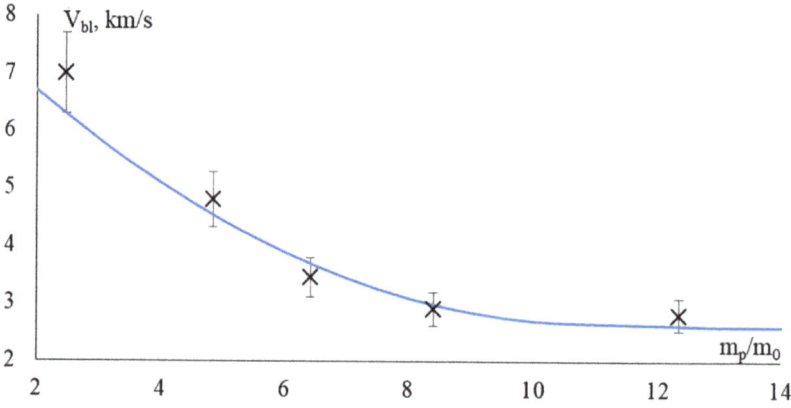

Figure 14. Dependence of the maximum ballistic velocity on the mass of the incoming particle according to the Formula (14). Crosses represent the results of numerical calculations.

Let us perform the calculations for the sequential collision of a group of eight particles with the protective screens, as shown in Figure 5. Random values close to 4 μs were taken as time intervals, and the particles were spatially located in a circle of 0.5 cm radius with the first particle at the center, as is indicated in Figure 12. The most resistant screens were selected for testing, namely, screen #5, which is a composite with volumetric reinforcement; screen #4, which is a layered screen of A356/316L; and screen #7, which is a heterogeneous screen with a direct gradient of 316L steel on an A356 matrix.

The calculation results are presented in Figure 15, which shows cross-sections of protective screens with a thickness of approximately one step of the computational grid cut out from the central region. The sequential multiple impact loading by SD particles leads to the propagation of shock waves through the deformed and partially damaged materials of the screens, resulting in further damage and fracture of the materials. The common feature of all screens is the process of damage and fracture of the A356 aluminum alloy, which is quite brittle. For the volumetrically reinforced composite, delamination between the steel elements and the aluminum matrix is characteristic, as can be seen in Figure 15a, and individual grains appear on the back side of the screen due to spallation processes. The mass of such grains is 0.1 mg, and their velocity is 50–70 m/s, so they do not pose a significant danger to the protected devices.

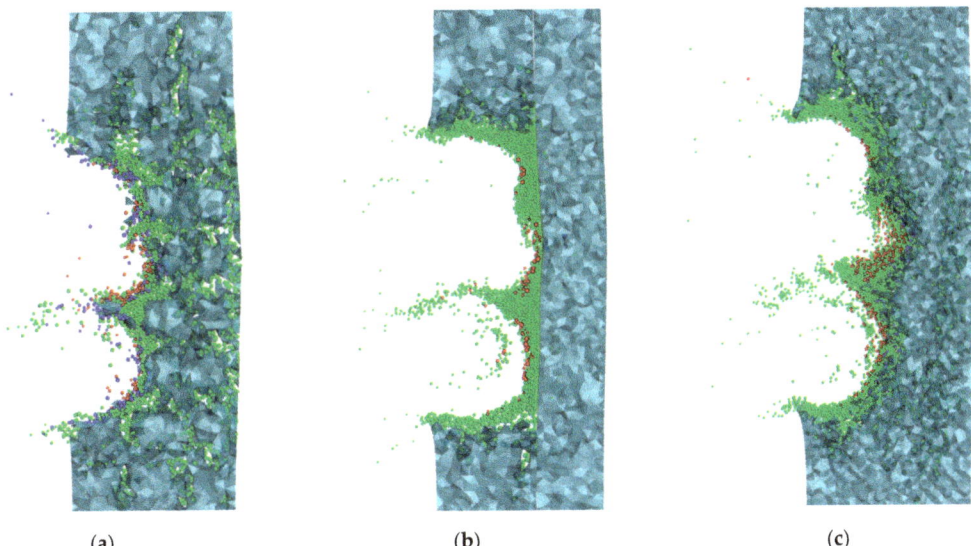

Figure 15. Cross-sectional view of the screens after the impact of a 8 SD particle flow: (**a**) screen #5; (**b**) screen #4; (**c**) screen #7.

The two-layer A356/316L screen withstands a similar impact load because the back side of the screen is made of 316L steel, which has sufficient strength against the spallation. In this case, the size of the crater volume is the largest among the compared protective screens since the A356 aluminum alloy quickly becomes damaged and starts to break down, as shown in Figure 15b. There are no detached microscopic grains behind the barrier since the back side is made of steel.

A heterogeneous screen with a direct gradient of steel has sufficient resistance to the particle flow impact. Since the front side of the screen is made of pure A356 alloy, it is natural that the initial stage of screen deformation resembles that of a layered screen. However, as the process advances into the depth of the screen, the resistance to deformation increases due to the inclusion of 316L steel. Therefore, the main damages and fractures of the screen occur on the front side, and the exit of waves to the back side of the screen does not lead to spall phenomena since the back side of the screen consists of pure 316L steel. However, the damage to the aluminum matrix is present throughout the volume of the screen (see Figure 15c).

As all three screens have practically equal chances of protecting an important element of the spacecraft from both a single SD particle and a particle stream, let us consider the capability of the protective screens to withstand a large SD particle twice the diameter at the same impact velocity. Thus, the mass of the impacting SD particle equals eight masses of the particle used in the experiments and calculations described above. Figure 16 shows the comparison of the results of the calculations of the impact of a large SD particle on the screens. All selected screens have through holes.

Screen #5, made of the volumetrically reinforced composite, has a hole close to a cylindrical shape. Large fragments are the fragments of reinforcing steel, while small fragments are the fragments of the aluminum matrix. The velocity of the large fragments in the head part of the barrier stream reaches 380–400 m/s, while that of the small ones reaches 450–475 m/s. On the periphery of the through cavity, stratifications of the reinforcement and matrix are noticeable due to the large deformations (see Figure 16a).

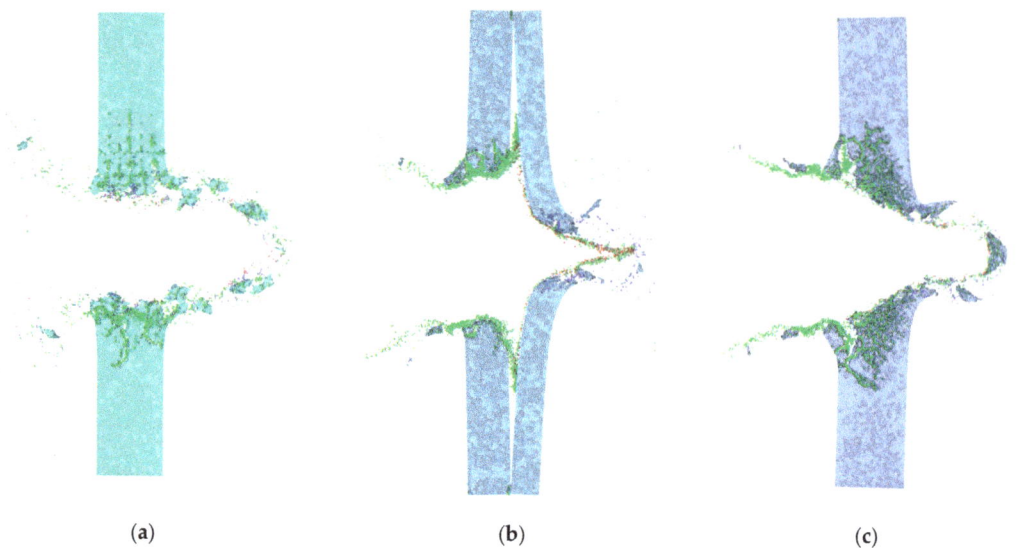

(a) (b) (c)

Figure 16. The result of numerical modeling for the process of large high-speed SD particles penetrating the screens: (**a**) screen #5, (**b**) screen #4, (**c**) screen #7.

The hole in the two-layer screen #4 has a more complex structure since the front aluminum plate received the main energy of the impacting SD particle, which led to the significant deformation of the material and subsequent damage and fracture. The rear steel plate of the screen was subjected to a weakened impact, which resulted in its bending. The stratification of the plates also allowed the shattered aluminum fragments to move radially, reducing the impact on the steel plate, so the final hole had small dimensions and was only in the central area. The steel damage was present in a small vicinity of the through hole. The fragment velocities in the cloud behind the screen were the same as above (see Figure 16b).

Screen #7 is the heterogeneous screen with a direct distribution of steel in the aluminum matrix and also has a through hole due to the impact of a large SD particle. The front side of the screen suffered significant damage due to the release of strong compression waves on the periphery of the initially formed crater, resulting in the detachment of sufficiently thick layers. Since the screen strength increases with the depth of penetration, a layer with large deflections is formed near the rear surface, and the cavity narrows. Further, the residual mass of the impacting SD particle breaks through the thin steel layer, forming a plug. The velocity of both large and small fragments is approximately 360–380 m/s (see Figure 16c).

9. Conclusions

The development and advancement of technologies for creating materials with specific characteristics, such as additive manufacturing, have broadened the range of applications for complex heterogeneous metal matrix materials. Experimental work on the practical implementation of production technologies for such heterogeneous media surpasses the number of studies on material prediction methods and the level of understanding of their properties under intense dynamic loads. This results in a significant gap when it comes to the practical utilization of heterogeneous materials with predetermined properties.

In our study, a significant stride has been made in comprehending the design of heterogeneous materials through direct numerical modeling of deformation and failure processes in such materials under high-speed loading. Furthermore, we have demonstrated that:

1. For the first time, taking into account the fracture effects, a numerical solution has been obtained for the problem of high-velocity interaction between space debris

particles and a volumetrically reinforced penetrating composite screen. It has been demonstrated that the screens constructed as two-layer A356/316L screens, volume-reinforced composites, and heterogeneous screens with a direct gradient distribution of steel in an aluminum matrix provide protection to devices from both individual space debris particles and streams of debris particles moving at speeds up to 6.0 km/s.

2. The physico-mechanical parameters of the heterogeneous material behind the shock wave front, obtained through numerical calculations, show deviations from the parameters calculated using the mixture rule of no more than 3%.

3. It has been shown that reinforcing the aluminum matrix with discrete steel inclusions within the specified mass and dimensional parameters of the screens does not provide sufficient protection for spacecraft components against high-velocity space debris particles.

Author Contributions: Conceptualization, I.S. and E.K.; software, E.K.; validation, A.K. and I.S.; writing—original draft preparation, A.K. and I.S.; writing—review and editing, A.B., I.S. and E.K.; supervision, E.K.; project administration, A.B.; funding acquisition, E.K. All authors have read and agreed to the published version of the manuscript.

Funding: This research was funded by Russian Science Foundation, project number 23-29-00777, available at https://rscf.ru/en/project/23-29-00777/ (accessed on 1 June 2023).

Institutional Review Board Statement: Not applicable.

Informed Consent Statement: Not applicable.

Data Availability Statement: Not applicable.

Conflicts of Interest: The authors declare no conflict of interest.

References

1. Smirnov, N.N. *Space Debris*; Smirnov, N.N., Ed.; CRC Press: London, UK, 2001; ISBN 9781482288193.
2. Anz-Meador, P.D.; Opiela, J.N.; Shoots, D.; Liou, J.-C. *History of On-Orbit Satellite Fragmentations*, 15th ed.; NASA: Houston, TX, USA, 2018.
3. Whipple, F.L. Meteorites and Space Travel. *Astron. J.* **1947**, *52*, 131. [CrossRef]
4. Christiansen, E.L.; Crews, J.L.; Williamsen, J.E.; Robinson, J.H.; Nolen, A.M. Enhanced Meteoroid and Orbital Debris Shielding. *Int. J. Impact Eng.* **1995**, *17*, 217–228. [CrossRef]
5. Zhuang, S.; Ravichandran, G.; Grady, D.E. An Experimental Investigation of Shock Wave Propagation in Periodically Layered Composites. *J. Mech. Phys. Solids* **2003**, *51*, 245–265. [CrossRef]
6. Grujicic, M.; Pandurangan, B.; Bell, W.C.; Bagheri, S. Shock-Wave Attenuation and Energy-Dissipation Potential of Granular Materials. *J. Mater. Eng. Perform.* **2012**, *21*, 167–179. [CrossRef]
7. Lamberson, L. Investigations of High Performance Fiberglass Impact Using a Combustionless Two-Stage Light-Gas Gun. *Procedia Eng.* **2015**, *103*, 341–348. [CrossRef]
8. Rawal, S.P. Metal-Matrix Composites for Space Applications. *JOM* **2001**, *53*, 14–17. [CrossRef]
9. Kok, Y.; Tan, X.P.; Wang, P.; Nai, M.L.S.; Loh, N.H.; Liu, E.; Tor, S.B. Anisotropy and Heterogeneity of Microstructure and Mechanical Properties in Metal Additive Manufacturing: A Critical Review. *Mater. Des.* **2018**, *139*, 565–586. [CrossRef]
10. DebRoy, T.; Wei, H.L.; Zuback, J.S.; Mukherjee, T.; Elmer, J.W.; Milewski, J.O.; Beese, A.M.; Wilson-Heid, A.; De, A.; Zhang, W. Additive Manufacturing of Metallic Components–Process, Structure and Properties. *Prog. Mater. Sci.* **2018**, *92*, 112–224. [CrossRef]
11. Filippov, A.A.; Fomin, V.M.; Buzyurkin, A.E.; Kosarev, V.F.; Malikov, A.G.; Orishich, A.M.; Ryashin, N.S. The Development of Heterogeneous Materials Based on Ni and B4C Powders Using a Cold Spray and Stratified Selective Laser Melting Technologies. *J. Phys. Conf. Ser.* **2018**, *946*, 012005. [CrossRef]
12. Fomin, V.M.; Golyshev, A.A.; Malikov, A.G.; Orishich, A.M.; Filippov, A.A. Creation of a functionally gradient material by the selective laser melting method. *J. Appl. Mech. Tech. Phys.* **2020**, *61*, 878–887. [CrossRef]
13. Fomin, V.M.; Golyshev, A.A.; Kosarev, V.F.; Malikov, A.G.; Orishich, A.M.; Filippov, A.A. Deposition of Cermet Coatings on the Basis of Ti, Ni, WC, and B4C by Cold Gas Dynamic Spraying with Subsequent Laser Irradiation. *Phys. Mesomech.* **2020**, *23*, 291–300. [CrossRef]
14. Zhang, M.; Yu, Q.; Liu, Z.; Zhang, J.; Tan, G.; Jiao, D.; Zhu, W.; Li, S.; Zhang, Z.; Yang, R.; et al. 3D Printed Mg-NiTi Interpenetrating-Phase Composites with High Strength, Damping Capacity, and Energy Absorption Efficiency. *Sci. Adv.* **2020**, *6*, eaba5581. [CrossRef]
15. Pawlowski, A.E.; Splitter, D.A.; Muth, T.R.; Shya, A.; Carver, J.K.; Dinwiddie, R.B.; Elliott, A.M. Producing Hybrid Metal Composites by Combining Additive Manufacturing and Casting. *Adv. Mater. Process.* **2017**, *175*, 16–21.

16. Pawlowski, A.E.; Cordero, Z.C.; French, M.R.; Muth, T.R.; Keith Carver, J.; Dinwiddie, R.B.; Elliott, A.M.; Shyam, A.; Splitter, D.A. Damage-Tolerant Metallic Composites via Melt Infiltration of Additively Manufactured Preforms. *Mater. Des.* **2017**, *127*, 346–351. [CrossRef]
17. Moustafa, A.R.; Dinwiddie, R.B.; Pawlowski, A.E.; Splitter, D.A.; Shyam, A.; Cordero, Z.C. Mesostructure and Porosity Effects on the Thermal Conductivity of Additively Manufactured Interpenetrating Phase Composites. *Addit. Manuf.* **2018**, *22*, 223–229. [CrossRef]
18. Poole, L.L.; Gonzales, M.; French, M.R.; Yarberry, W.A.; Moustafa, A.R.; Cordero, Z.C. Hypervelocity Impact of PrintCast 316L/A356 Composites. *Int. J. Impact Eng.* **2020**, *136*, 103407. [CrossRef]
19. French, M.R.; Yarberry, W.A., III; Pawlowski, A.E.; Shyam, A.; Splitter, D.A.; Elliott, A.M.; Carver, J.K.; Cordero, Z.C. Hypervelocity Impact of Additively Manufactured A356/316L Interpenetrating Phase Composites. In Proceedings of the Solid Freeform Fabrication 2017: Proceedings of the 28th Annual International Solid Freeform Fabrication Symposium—An Additive Manufacturing Conference Hypervelocity, Austin, TX, USA, 7–9 August 2017.
20. Cheng, J.; Gussev, M.; Allen, J.; Hu, X.; Moustafa, A.R.; Splitter, D.A.; Shyam, A. Deformation and Failure of PrintCast A356/316L Composites: Digital Image Correlation and Finite Element Modeling. *Mater. Des.* **2020**, *195*, 109061. [CrossRef]
21. Kraus, A.E.; Kraus, E.I.; Shabalin, I.I. Numerical Simulation of the High-Speed Interaction of a Spherical Impactor with a System of Spaced Heterogeneous Plates. *J. Phys. Conf. Ser.* **2019**, *1404*, 012026. [CrossRef]
22. Kraus, A.E.; Kraus, E.I.; Shabalin, I.I. Impact Resistance of Ceramics in a Numerical Experiment. *J. Appl. Mech. Tech. Phys.* **2020**, *61*, 847–854. [CrossRef]
23. Kraus, A.E.; Kraus, E.I.; Shabalin, I.I.; Buzyurkin, A.E. Evolution of a Short Compression Pulse in a Heterogeneous Elastoplastic Medium. *J. Appl. Mech. Tech. Phys.* **2021**, *62*, 475–483. [CrossRef]
24. Kraus, A.E.; Kraus, E.I.; Shabalin, I.I. Reactor 3D Software Performance on Penetration and Perforation Problems. In *Behavior of Materials under Impact, Explosion, High Pressures and Dynamic Strain Rates*; Orlov, M.Y., Visakh, P.M., Eds.; Springer International Publishing: Berlin/Heidelberg, Germany, 2023; pp. 83–101.
25. Fomin, V.M.; Gulidov, A.I.; Sapozhnikov, G.A.; Shabalin, I.I. *High-Velocity Solids Interaction*; SB RAS: Novosibirsk, Russia, 1999; ISBN 5-7692-0237-8.
26. Wilkins, M.L. *Computer Simulation of Dynamic Phenomena*; Scientific Computation; Springer: Berlin/Heidelberg, Germany, 1999; ISBN 978-3-642-08315-0.
27. Kraus, E.I.; Shabalin, I.I. A Few-Parameter Equation of State of the Condensed Matter. *J. Phys. Conf. Ser.* **2016**, *774*, 012009. [CrossRef]
28. Kraus, E.; Shabalin, I. Melting behind the Front of the Shock Wave. *Therm. Sci.* **2019**, *23*, 519–524. [CrossRef]
29. Maenchen, G.; Sack, E. The Tensor Code. In *Methods in Computational Physics, v.3, Fundamental Methods in Hydrodynamics*; Alder, B., Fernbach, S., Rotenberg, M., Eds.; Academic Press: New York, NY, USA, 1964; p. 221.
30. Tuler, F.R.; Butcher, B.M. A Criterion for the Time Dependence of Dynamic Fracture. *Int. J. Fract. Mech.* **1968**, *4*, 322–328. [CrossRef]
31. Johnson, G.R.; Stryk, R.A. Conversion of 3D Distorted Elements into Meshless Particles during Dynamic Deformation. *Int. J. Impact Eng.* **2003**, *28*, 947–966. [CrossRef]
32. Kraus, E.I.; Shabalin, I.I. A New Model to Determine the Shear Modulus and Poisson's Ratio of Shock-Compressed Metals up to the Melting Point. *High Press. Res.* **2021**, *41*, 353–365. [CrossRef]
33. Kraus, E.I.; Shabalin, I.I.; Shabalin, T.I. Numerical Analysis of Wave Propagation in a Cermet Composite. In *AIP Conference Proceedings*; AIP Publishing LLC: New York, NY, USA, 2017; Volume 1893, p. 030130.
34. Kraus, A.E.; Kraus, E.I.; Shabalin, I.I. Simulation of a Group Impact on a Heterogeneous Target of Finite Thickness. *J. Sib. Fed. Univ. Math. Phys.* **2021**, *14*, 1–12. [CrossRef]
35. Kraus, A.E.; Kraus, E.I.; Shabalin, I.I. A Heterogeneous Medium Model and Its Application in a Target Perforation Problems. In *Multiscale Solid Mechanics. Advanced Structured Materials*; Altenbach, H., Eremeyev, V.A., Igumnov, L.A., Eds.; Springer: Berlin/Heidelberg, Germany, 2021; Volume 141, pp. 289–304.
36. Hashin, Z. The Elastic Moduli of Heterogeneous Materials. *J. Appl. Mech.* **1962**, *29*, 143–150. [CrossRef]
37. Hashin, Z.; Rosen, B.W. The Elastic Moduli of Fiber-Reinforced Materials. *J. Appl. Mech.* **1964**, *31*, 223–232. [CrossRef]
38. Dremin, A.N.; Karpukhin, I.A. Method of Determining the Shock Adiabats for Disperse Materials. *Zh. Prikl. Mekh. I Tekh. Fiz.* **1960**, *1*, 184–188.
39. Duvall, G.E.; Taylor, S.M. Shock Parameters in a Two Component Mixture. *J. Compos. Mater.* **1971**, *5*, 130–139. [CrossRef]
40. Bakhvalov, N.; Panasenko, G. *Homogenisation: Averaging Processes in Periodic Media*; Mathematics and its Applications; Springer: Dordrecht, The Netherlands, 1989; Volume 36, ISBN 978-94-010-7506-0.
41. Christensen, R.M. *Mechanics of Composite Materials*; Dover Publications: Mineola, NY, USA, 2005; ISBN 978-0486442396.
42. Poole, L.L.; Gonzales, M.; Moustafa, A.R.; Gerlt, A.R.C.C.; Cordero, Z.C. Shock Dynamics in Periodic Two-Dimensional Composites. In *AIP Conference Proceedings*; American Institute of Physics Inc.: Melville, NY, USA, 2020; Volume 2272, p. 120020.
43. Taylor, S.V.; Gonzales, M.; Cordero, Z.C. Shock Response of Periodic Interpenetrating Phase Composites. *APL Mater.* **2022**, *10*, 111119. [CrossRef]
44. Baluch, A.H.; Park, Y.; Kim, C.G. High Velocity Impact Characterization of Al Alloys for Oblique Impacts. *Acta Astronaut.* **2014**, *105*, 128–135. [CrossRef]

45. Kim, Y.H.; Lee, S.; Kim, N.J.; Cho, K.M. Effect of Microstructure on the Tensile and Fracture Behavior of Cast A356 AlSiCp Composite. *Scr. Metall. Mater.* **1994**, *31*, 1629–1634. [CrossRef]
46. Hixson, R.S.; McQueen, R.G.; Fritz, J.N. The Shock Hugoniot of 316 SS and Sound Velocity Measurements. In *AIP Conference Proceedings*; AIP: Melville, NY, USA, 1994; Volume 309, pp. 105–108.
47. Taylor, G.I. The Formation and Enlargement of a Circular Hole in a Thin Plastic Sheet. *Q. J. Mech. Appl. Math.* **1948**, *1*, 103–124. [CrossRef]

Disclaimer/Publisher's Note: The statements, opinions and data contained in all publications are solely those of the individual author(s) and contributor(s) and not of MDPI and/or the editor(s). MDPI and/or the editor(s) disclaim responsibility for any injury to people or property resulting from any ideas, methods, instructions or products referred to in the content.

Article

Extending the NNO Ballistic Limit Equation to Foam-Filled Dual-Wall Systems

William P. Schonberg

Civil Engineering Department, Missouri University of Science & Technology, Rolla, MO 65409, USA; wschon@mst.edu

Abstract: A key component in the quantitative assessment of the risk posed to spacecraft by the micrometeoroid and orbital debris (MMOD) environment is frequently referred to as a ballistic limit equation (BLE). A frequently used BLE for dual-wall configurations (which are commonly used on spacecraft to protect them against the MMOD environment) is the New Non-Optimum, or "NNO", BLE. In design applications where a BLE is needed for a new structural system that has not yet been tested, but resembles to a fair degree a dual-wall system, it is common practice to equivalence the materials, thicknesses, etc., of the new system to the materials, thicknesses, etc., of a dual-wall system. In this manner, the NNO BLE can be used to estimate the failure / non-failure response characteristics for the new system. One such structural wall system for which a BLE does not yet exist is a dual-wall system that is stuffed with a lightweight polymer-based foam material. In this paper we demonstrate that the NNO BLE, in its original form, frequently over- or under-predicts the response of such a system. However, when the NNO BLE is modified to more properly include the effects of the presence of the foam as well as the actual material properties of the walls and the impacting projectile, there is a marked improvement in its predictive abilities.

Keywords: ballistic limit equation; dual-wall system; foam-filled; hypervelocity impact; space debris

1. Introduction

A key component in a probabilistic risk assessment for spacecraft being designed to operate in the micrometeoroid and orbital debris (MMOD) environment is a ballistic limit equation, or BLE. This is an equation used to determine if a spacecraft component will suffer a critical failure following an on-orbit high-speed impact.

One type of BLE is derived from a damage predictor equation that is itself obtained from a curve-fit of a damage measurements, such as crater depth or hole diameter, in terms of impact parameters, material properties, and target configuration. The other kind of BLE is basically a "hand-drawn" discriminant line that, for example, separates (projectile diameter, impact velocity) combinations that cause failure from those that do not. It is these types of BLEs that were used to design the MMOD shielding on the International Space Station (see, e.g., [1] for more information on how such BLEs were developed).

A frequently used BLE for dual-wall configurations (also known as "Whipple Shields") is the New Non-Optimum, or NNO, BLE [2]. These wall designs are frequently used on spacecraft to protect them against the threats posed by MMOD particles impacts. The NNO BLE consists of three parts in terms of increasing impact velocity—a low velocity portion (through approx. 3 km/s), a high velocity portion (above approx. 7 km/s), and a linear interpolation between the BLE values at the end of and the start of the low and high velocity portions, respectively.

Quite frequently, a BLE is needed for a new structural system or element that has not yet been tested, but resembles to a fair degree a dual-wall system for which the NNO BLE would be applicable. It such cases it is common practice to equivalence the materials, thicknesses, etc., of the system or element of interest (but for which a BLE does not yet

Citation: Schonberg, W.P. Extending the NNO Ballistic Limit Equation to Foam-Filled Dual-Wall Systems. *Appl. Sci.* **2023**, *13*, 800. https://doi.org/10.3390/app13020800

Academic Editors: Lorenzo Olivieri, Kanjuro Makihara and Leonardo Barilaro

Received: 15 November 2022
Revised: 28 December 2022
Accepted: 29 December 2022
Published: 6 January 2023

Copyright: © 2023 by the author. Licensee MDPI, Basel, Switzerland. This article is an open access article distributed under the terms and conditions of the Creative Commons Attribution (CC BY) license (https://creativecommons.org/licenses/by/4.0/).

exist) to the materials, thicknesses, etc., of a dual-wall system. In this manner, the NNO BLE can be used to estimate the failure/non-failure response characteristics for the new system of interest without having to expend significant resources to generate a BLE for that new system.

One such structural wall system for which a BLE does not yet exist, but is also seeing an increase in application, is a dual-wall system that is completely filled with a lightweight foam material (i.e., there are no discernable air gaps between either wall of the dual-wall system and the foam filling the space between them). The outer and inner walls in such a system could be aluminum, or made out of a composite material. Since such a system bears a close resemblance to a more standard dual-wall system (where the space between the outer and inner walls is empty), it could be considered appropriate to re-cast the foam-stuffed dual-wall system as an "empty" all-aluminum Whipple Shield with wall thicknesses that take into account any non-aluminum wall materials as well as the presence of the foam stuffing in the original system.

In such a dual-wall configuration, the foam between the outer and inner walls can be either metallic (e.g., lightweight aluminum foams), or non-metallic (e.g., lightweight polymer foams like polyurethane). The focus of the study described herein was on lightweight non-metallic polymer-based foams. Even then, the space between the outer and inner walls could be either fully filled or partially filled. We again focus our attention on dual-wall systems that are fully filled with foam. In this manner, the foam in the configuration we studied not only could affect the protective capability of the dual-wall system, but it also would provide some structural support to keep the outer wall at a constant distance away from the inner wall. This is especially important in applications where the outer and inner walls might be made of extremely flexible materials, such as composite fabrics or very thin metallic plates.

As will be seen shortly, we found that the NNO BLE in its original form frequently over- or under-predicted the response of such a system. However, when the NNO BLE was modified to more properly include the effects of the presence of the stuffing as well as the actual material properties of the walls and the impacting projectile, we found that there was a marked improvement in its predictive abilities.

In this paper, then, we present the results of a study whose goal was to improve the predictive ability of the NNO BLE when it is applied to the particular dual-wall construction involving metallic or non-metallic outer and inner walls, the space between which is filled with a non-metallic lightweight foam. In this study, we developed a set of functions that, when incorporated into the NNO BLE, does significantly improve its predictive ability for this type of wall system. This is demonstrated by comparing the predictions of the original and modified versions of the NNO BLE against experimental data and the results of hydrocode simulations for a variety of wall materials, foam materials, and projectile materials, and for impact velocities ranging from approx. 2–40 km/s.

2. Impact Conditions and Dual-Wall Constructions

A sketch of the target is shown in Figure 1—it consists of outer and inner walls (which could be either metallic or non-metallic) separated from each other by a small gap that is filled with a lightweight (or low density) non-metallic foam.

The Dual-wall Configuration

Figure 1. Sketch of a Dual-Wall Configuration Stuffed with Light-weight Foam.

Tables 1–3 below present a summary of the impact parameters, configurations, and material properties of the dual-wall targets used in the experimental test programs and numerical simulations that generated the data used in this study. Projectile density values in Table 1 were found, for the most part, in the reports or articles summarizing the test programs wherein those projectiles were used; others, where values were not provided, were obtained from an online database [3]. Additionally, the "diameters" of the disk projectiles are actually the equivalent spherical projectile diameters calculated using an equal-mass consideration.

Table 1. Impact Conditions in Previous Experimental Programs and Numerical Simulations.

Ref #	Projectile Shape	Projectile Material	Proj Mat'l Density (gm/cm³)	Proj Diam (mm)	Impact Velocity (km/s)	Impact Obliquity (deg)
[4]	Sphere	Pyrex	2.12	1.60	5.8–6.5	0
[5]	Disk	MgLi	1.35	0.909	5.0–5.5	0
[6]	Sphere	Aluminum	2.80	6.35	5.0–6.0	0
[6]	Disk	Lexan	1.20	6.60, 7.27	4.5–8.2	0
[7]	Disk	PETP [1]	1.38	2.49	2.0–5.8	0
[8–11]	Sphere	Al 2017-T4	2.80	1.9–7.0	6.8–7.1	0, 30, 60
[12]	Sphere	Al 2017-T4	2.80	0.8–2.1	7, 25	30

[1] PETP . . . polyethylene terephthalate.

Table 2. Target Configurations and Materials Used in Previous Experimental Programs and Numerical Simulations.

Ref #	Outer Wall Material	Outer Wall Thick (cm)	Filler Material	Filler Material Density (gm/cm³)	Filler Material Thick (cm)	Inner Wall Material	Inner Wall Thick (cm)
[4]	Al 2024-T3	0.030–0.056	Polyurethane	0.005–0.102	5.08, 7.62	Al 2024-T3	0.030–0.056
[5]	Al 2024-T3	0.0076	Polyurethane, Styrofoam	0.0285, 0.0288	0.483	Al 2024-T3	0.0127, 0.0254
[6]	Al 2024-T3	0.0508	Polyurethane	0.0320, 0.0336	3.81	Al 2024-T3	0.127
[7]	PETP [1] Fabric	0.127	Polyurethane	0.0230	3.94–4.32	PETP [1] (Coated and Uncoated)	0.063–0.127
[7]	Laminated and Unlaminated Rayon Fabric	0.064, 0.089	Polyurethane	0.0208	3.81–4.18	Rayon Fabric	0.071
[8–10]	Al 6061-T6	0.05	Polyimide	0.0056	2.0	Al 6061-T6	0.05
[11]	T300/Epoxy	0.097	Polymethacrylimide	0.0521	2.35	T300/Epoxy	0.097
[12]	Al 6061-T6	0.05	Polyimide	0.0056	2.0	Al 6061-T6	0.05
[12]	Glass/Epoxy	0.05	Polyimide	0.0056	5.0	IM7/Epoxy	0.101

[1] PETP . . . polyethylene terephthalate.

Filler material density values in Table 2 are, except for Refs. [8–12], as specified in the reports or articles summarizing the test programs wherein those materials were used; density values for polyimide and polymethacrylimide were also obtained from the same online database [3]. Likewise, in Table 3, outer and inner wall density values are also, for the most part, as specified in the various referenced reports or articles; density values not provided in the test program references were either obtained from the online database [3], or calculated using other information given elsewhere in the referenced documents. Outer and inner wall strength values were typically not provided in the reference documents, and so were estimated using strength values for similar materials provided elsewhere as indicated.

Table 3. Density and Strength Values for Inner Wall Materials.

Wall Type	Material	Density (gm/cm³)	Source	Strength (ksi)	Source
Outer Wall	PETP Fabric	0.769	Values as specified in references where used matweb.com	Outer wall strength values not required for NNO BLE	
	Rayon Fabric	0.713 [1], 0.474 [2]			
	T300/Epoxy	1.53			
	Glass/Epoxy	1.90			
Inner Wall	PETP Fabric	0.953 [3], 0.832 [4]	Values as specified in references where used Calculated	56.8	Ref. [13]
	Rayon Fabric	0.549		60.0	Ref. [14]
	T300/Epoxy	1.53, 1.64		290, 264	Toray data sheets
	IM7/Epoxy	1.58		397	Hexcel data sheets

[1] Laminated Fabric, [2] Unlaminated fabric, [3] Elastomer coated fabric, [4] Uncoated fabric.

3. Modifications to the NNO BLE

In this study we developed a set of functions that, when incorporated into the NNO BLE, significantly improve its predictive ability for a foam-filled dual-wall system. These functions were intended to more properly take into account the material properties of the impacting projectile and the walls in the dual-wall system. In the modified NNO BLE, the presence of the foam in the dual-wall system under consideration is taken into account in the same manner in which they are traditionally included when the original NNO BLE is applied to such dual-wall systems. Namely, in both cases, the thicknesses of the outer and inner walls are increased slightly using a mass equivalence calculation that allocated 50% of the foam filler mass to the outer wall and 50% to the inner wall.

The following equation gives a top-level perspective of how the original NNO BLE is to be modified for these types of wall configurations:

$$d_{crit}^{mod} = d_{crit}^{orig} * f_1(\rho_{rw}, \sigma_{rw}) * f_2(\rho_p) * f_3(\theta_p) \tag{1}$$

where d_{crit}^{mod} and d_{crit}^{orig} are the modified and original critical, or ballistic limit, projectile diameters as predicted by the modified and original NNO BLE, respectively. In this equation, the function f_1 accounts for the inner wall density and strength, if different from aluminum, the function f_2 accounts for the density of non-aluminum projectiles, and the function f_3 accounts for the effects of impact obliquity.

The forms of the modifier functions $f_1, f_2,$ and f_3 in Equation (1) are guided by expected asymptotic function values or the roles played by those functions in modifying the original NNO BLE. For example, as projectile density approaches that of aluminum (from below, that is, when ρ_p becomes greater than ~2.0 gm/cm³), all modification functions should approach unity (i.e., the modifiers should all approach unity when aluminum projectiles are considered because that is the projectile material on which–for the most part–the NNO BLE is based). The same should be true when aluminum walls are used, that is, when ρ_{rw} and σ_{rw} take on values corresponding to those of aluminum–in these cases, the values of the modifier functions should then also all approach unity. The following equations define the modifying functions $f_1, f_2,$ and f_3:

$$f_1(\rho_{rw}, \sigma_{rw}) = 1 - exp\left\{-15.73\left[(f_{RWS} * f_{RWD})^{-9.826}\right]\right\} \tag{2}$$

where

$$f_{RWS} = (\sigma_{RW}/50)^{\{1-exp[-7,158(\sigma_{RW}/50)^{-3.949}]\}} \tag{3a}$$

$$f_{RWD} = (\rho_{RW}/2.71)^{\{3.0[1-exp(-0.007967\rho_{RW}{}^{12.34})]\}} \tag{3b}$$

$$f_2(\rho_p) = 1 + (MF_0 - 1) * \left[1 - exp\left(-f_{PD}(1 - V_P/72)^{f_{VP}}\right)\right] / [1 - exp(-f_{PD})] \tag{4}$$

with

$$\begin{Bmatrix} MF_0 \\ f_{PD} \\ f_{VP} \end{Bmatrix} = 1 + (A-1)\left\{1 - exp\left[-B(1-\rho_P/10.0)^C\right]\right\}/[1-exp(-B)] \tag{5}$$

where the constants A, B, and C are given for each equation in Table 4 below, and

$$f_3(\theta_p) = 1 + 0.075 V_P^{0.9033} \theta_P^{0.2084} \tag{6}$$

In Equations (2), (3) and (6), constants and coefficients with 4 significant figures were obtained through curve-fitting exercises using the desired functional forms. As noted previously, these functional forms were informed by desired function values as inner wall material property values approached certain values within the impact test database used in this study.

Alternatively, the constants 50, 2.71, and 72 (and the 2.85 in Equation (5) as well) were used to, in effect, non-dimensionalize corresponding numerators to render the terms within the desired functions to have values of similar orders of magnitude. This, in turn, facilitated the regression exercise that yielded the other constants in the various functions.

Finally, the values of the constants A, B, and C in Equations (4) and (5), that is, those in Table 3, were obtained manually using two considerations. First, the function values had to approach expected asymptotic values, and second, the correctness of the BLE predictions was maximized to the highest extent possible. That is, BLE predictions of ballistic limit diameter were checked to ensure that they were, as often as possible, (1) larger than actual projectile diameters in tests where the inner walls were not perforated, and (2) smaller than actual projectile diameters when inner walls were perforated.

Table 4. Parameter Values for Lower Projectile Density Function.

	A	B	C
MF_0	3.0	1.4×10^3	45
f_{PD}	2.4×10^3	2.0×10^9	163
f_{VP}	100.0	3.4×10^4	70

In these equations, the various input parameters are defined as follows:
ρ_P is the projectile material density (in gm/cm^3)
ρ_{RW} is the inner wall material density (in gm/cm^3)
σ_{RW} is the inner wall material tensile strength (in ksi)
θ_P is the trajectory obliquity (radians)
V_P is the impact velocity (km/s)

4. Comparison with Test Data and Numerical Simulation Predictions

The next series of plots shows comparisons between the predictions of the original NNO BLE and NNO BLE as modified according to Equations (1)–(4). Two types of plots were used for comparison for different configurations:

- D_{proj}/D_{crit} vs. V_{imp}—When the ratio > 1, did the test result in an inner wall perforation for that particular impact velocity? Likewise, when the ratio < 1, did the test result in a non-perforation event?

- D_{crit} vs. V_{imp}—Are the tests with the inner wall perforations above the ballistic limit curse, and are those without inner wall perforation below the curve, as impact velocity is increased?

In these plots, the original NNO BLE predictions of D_{crit} are calculated with the following parameter modifications as necessary:

- *Projectile Density*—density of actual projectile material
- *Outer Wall Thickness*—mass equivalent aluminum thickness (assuming an aluminum density of 2.71 gm/cm^3) for original outer wall material and thickness and 50% of the foam stuffing
- *Outer Wall Density*—density of original aluminum outer wall material, or 2.71 gm/cm^3 if mass equivalent aluminum outer thickness is being used
- *Stand-off Distance or Spacing*—this is the thickness of the foam between the outer wall and the inner wall
- *Inner Wall Thickness*—mass equivalent aluminum thickness (assuming an aluminum density of 2.71 gm/cm^3) for original inner wall material and thickness and 50% of the foam stuffing
- *Inner Wall Density*—density of original aluminum inner wall material, or 2.71 gm/cm^3 if mass equivalent aluminum bumper thickness is being used
- *Inner Wall Yield Strength*—actual yield strength for aluminum inner wall materials; ultimate tensile strength for non-aluminum inner wall materials

Figures 2–4 show a comparison between the plots of the modified and original NNO BLEs for several different constructions of polyurethane-filled dual-wall systems impacted by non-aluminum projectiles. Additionally, shown are the experimental results from [6,8] regarding whether or not the inner walls of the dual-wall systems were perforated (P) or not (NP). In Figure 2, the original and modified BLEs shown were obtained using inner wall density, inner wall thickness, and filler thickness parameter values averaged across the various dual-wall constructions in [8].

It is evident from these plots that the original formulation of the NNO BLE, as implemented above, did not adequately model the P/NP response of those particular foam-filled dual-wall systems. That is, while most of the tests with the inner wall perforations (the hollow P datapoints) were above the original NNO ballistic limit curves as expected, those without inner wall perforation (the solid NP datapoints) were not below the original NNO BLE curves. However, when the modifications to the original NNO BLE were implemented as described above, the hollow P datapoints (for the most part) remained above the modified NNO ballistic limit curves, while the solid NP datapoints were now (for the most part) below the modified NNO BLE curves.

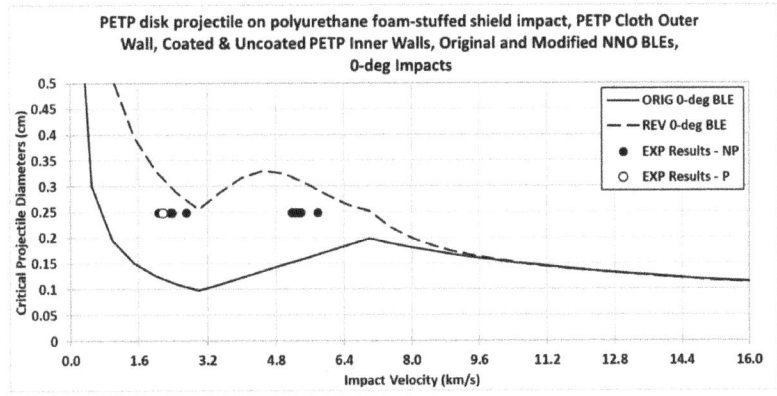

Figure 2. Comparison of Original and Modified NNO BLEs against P/NP data in [8] for Dual-Wall Systems with PETP Outer Walls and PETP Inner Walls.

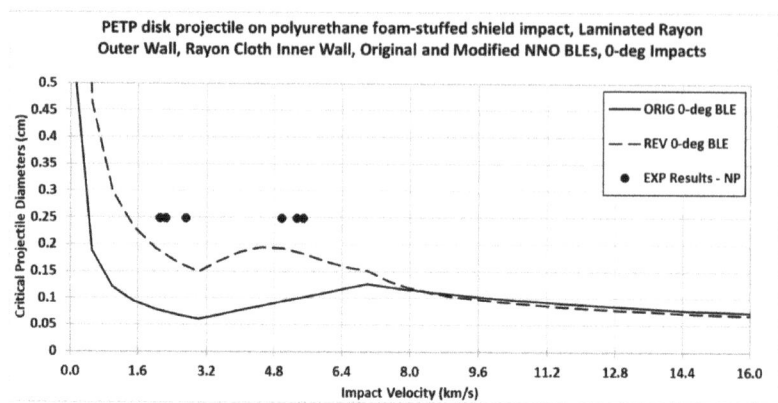

Figure 3. Comparison of Original and Modified NNO BLEs against P/NP data in [8] for Dual-Wall Systems with Rayon Outer and Inner Walls.

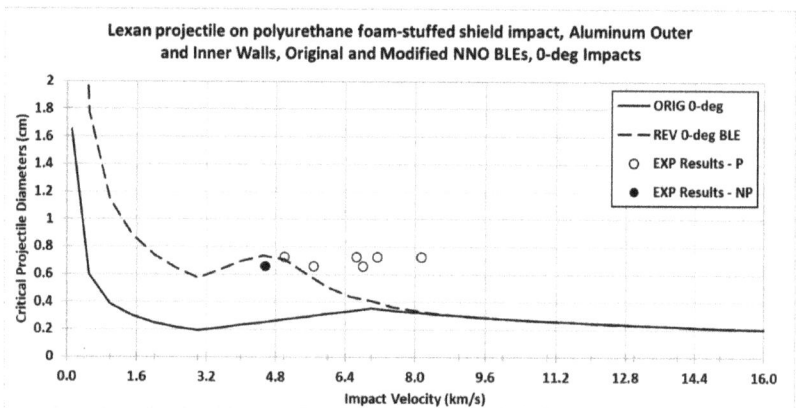

Figure 4. Comparison of Original and Modified NNO BLEs against P/NP data in [6] for Dual-Wall Systems with Aluminum Outer and Inner Walls.

This change is a significant improvement in the ability of the (modified) NNO BLE to predict the P/NP response of foam-filled dual-wall systems with aluminum as well as composite material outer and inner walls (at least for impact velocities between 2 and 8 km/s). Of course, as can be seen in Figure 3, the modified NNO BLE, while significantly closer to the NP datapoints than the original NNO BLE, failed to end up above those points. So while an accuracy improvement is still evident for the modified NNO BLE over the original NNO BLE for foam-filled dual-wall systems with laminated rayon outer walls and cloth or fabric inner walls, there are still some response characteristics not entirely correctly captured by the modifications made to the original NNO BLE for this particular dual-wall configuration.

In Figure 2, we also see that while all of the non-perforation datapoints are now below the modified NNO BLE, the perforation datapoint is not above it, as it should be. Of additional interest is that it appears to fall amidst a series of non-perforation datapoints, indicating that there might be something amiss with this test that resulted in a perforation. However, all of the datapoints in Figure 4 do appear to fall on the correct sides of the modified NNO BLE–the non-perforation datapoint is below it, and all of the perforation datapoints are above it.

Figures 5 and 6 confirm the ability of the modified NNO BLE to predict the P/NP response of foam-filled dual wall systems more correctly, but now also at velocities as high as 30 km/s. In these figures, the P/NP datapoints are not always where they might be expected to be found with respect to their placement about the original NNO BLE. In Figure 5, e.g., most of the NP points are above the plot of the original NNO BLE (whereas they should be below it), and in Figure 6, most of the P datapoints are below it, whereas they should be above it.

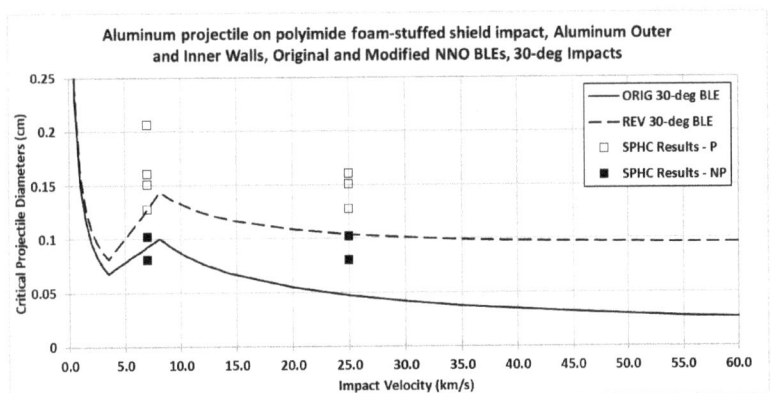

Figure 5. Comparison of Original and Modified NNO BLEs against P/NP data in [12] for Dual-Wall Systems with Aluminum Outer and Inner Walls.

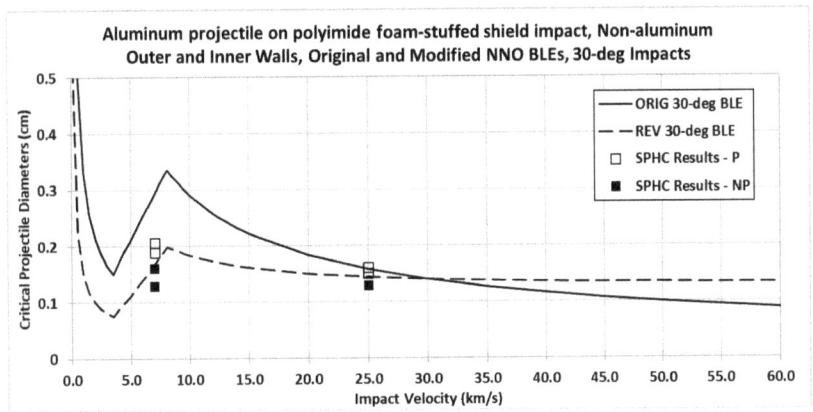

Figure 6. Comparison of Original and Modified NNO BLEs against P/NP data in [12] for Dual-Wall Systems with Non-Aluminum Outer and Inner Walls.

However, when the placements of the P/NP datapoints are compared against the plots of the modified NNO BLEs for these wall systems, we see that these points are now, for the most part, where they need to be. That is, the solid NP points in Figure 5 are now below the BLE curve, and in Figure 6, the hollow P points are now above it.

It is important to note that the comparisons shown in Figures 2–6 are those where the modified BLE is plotted against the datapoints used in the development of the modifications given by Equations (1)–(6). It would be instructive, of course, to compare the predictions of the original and modified NNO BLEs against some P/NP data from tests using targets that were not part of the dataset that was used in the development of those modifications. These comparisons are shown in Figures 7 and 8 below.

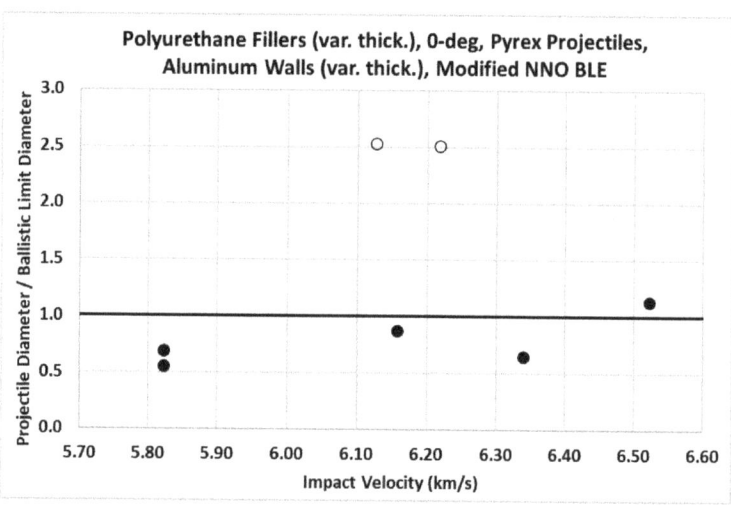

Figure 7. Comparison of Original and Modified NNO BLEs against P/NP data in [4,6] for Dual-Wall Systems with Aluminum Walls and a Polyurethane Filler.

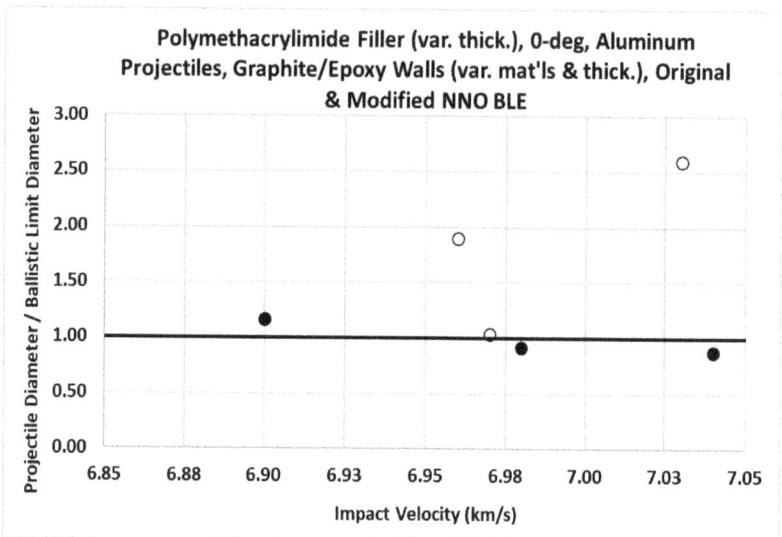

Figure 8. Comparison of Original and Modified NNO BLEs against P/NP data in [10,11] for Dual-Wall Systems with Non-Aluminum Walls and a Polymethacrylimide Filler.

In Figures 7 and 8 it is evident that the modified NNO BLE fared very well in predicting the P/NP response of the dual-wall systems and fillers under consideration. There was, of course, in each figure, one non-perforation datapoint that appeared on the side opposite to where it was expected. However, that kind of "spillage" is not at all surprising or unexpected (see, e.g., [15]).

Table 5 below presents a top-level overview of the ability of the original and modified NNO BLEs to correctly predict the P/NP response of a foam-filled dual-wall system. As can be seen in this table, there is a marked reduction in the number of incorrect response predictions (and a corresponding increase in the number of correct predictions) of the modified NNO BLE as compared to the original NNO BLE formulation. That is, in approx.

37% of the impact tests, the original NNO BLE incorrectly predicted whether or not the inner wall would be perforated. This occurred in only approx. 17% of the tests when the modified NNO BLE was used.

Table 5. Overview of Original and Modified NNO BLE Performance.

	Original NNO BLE		Modified NNO BLE	
P predicted as P	40	45%	42	47%
P predicted as NP	4	4%	1	1%
NP predicted as P	29	33%	14	16%
NP predicted as NP	16	18%	32	36%

As a final comment, we recall that the original formulation of the NNO BLE has three sections: a downward curving low velocity section, and upward sloping intermediate section, and a second less steep downward sloping high velocity section. In an effort to reduce the complexity of the modifications, the same modification formulation was applied to all three sections of the dual-wall BLE. The less-than-hoped for improvements to the NNO BLE when the modifications presented herein in the low and intermediate velocity sections seen in Figures 2, 3, 7 and 8 indicate that perhaps in might be necessary to investigate the possibility that the different sections of the dual-wall BLE might each need its own modification factor.

Similarly, the results in Figure 3 show that the predictions of the modified model at 0° are not as good as those at 30° as shown in Figures 5 and 6. The modification factor f_3 as given by Eq 6 does have an impact obliquity term. However, it might be necessary to explore the possibility of adding additional obliquity terms to the equations for modifications f_1 and f_2 to improve the agreement between experimental results and the modified NNO BLE, especially for 0-deg impacts.

5. Summary and Conclusions

A study was performed in which a frequently used BLE for dual-wall configurations that are commonly used on spacecraft to protect them against the MMOD environment was modified to more properly include the effects of any stuffing between the walls as well as the actual material properties of the walls and the impacting projectile. By comparing the predictions for ballistic limit diameter of the modified and original versions of this BLE, we found that there was, overall, a marked improvement in the response prediction ability once the modifications were introduced into the NNO BLE. This improvement was also evident when comparing original and modified NNO BLE predictions against empirical response data that were not used in developing these modifications. Using this modified version of the NNO BLE for these kinds of dual-wall systems will result in assessed mission risk values that would be more reflective of the actual spacecraft wall designs being used.

Funding: This research was supported by funding from the NASA Engineering Safety Center.

Informed Consent Statement: Not applicable.

Data Availability Statement: The data used in this study are wholly available in this article.

Acknowledgments: The author wishes to acknowledge the support provided by the NASA Engineering Safety Center that made this study possible. The author is also grateful to Joel Williamsen, Bob Stellingwerf, and Steve Evans for performing the hydrocode simulations that were used in this study, and to Eric Christiansen and the NASA team that performed the recent tests that were part of this study.

Conflicts of Interest: The author declares no conflict of interest.

References

1. Schonberg, W.P. Concise History of Ballistic Limit Equations for Multi-Wall Spacecraft Shielding. *Rev. Hum. Space Explor.* **2016**, *1*, 46–54. [CrossRef]
2. Christiansen, E.L. Design and Performance Equations for Advanced Meteoroid and Debris Shields. *Int. J. Impact Eng.* **1993**, *14*, 145–156. [CrossRef]
3. Material Property Data. Available online: https://matweb.com/ (accessed on 27 December 2022).
4. Pipitone, S.J.; Reynolds, B.W. Effectiveness of Foam Structures for Meteoroid Protection. *J. Spacecr.* **1964**, *1*, 37–43. [CrossRef]
5. Lundeberg, J.F.; Stern, P.H.; Bristow, R.J. *Meteoroid Protection for Spacecraft Structures*; NASA CR-554201; NASA: Washington, DC, USA, October 1965.
6. Lundeberg, J.F.; Lee, D.H.; Burch, G.T. Impact Penetration of Manned Space Stations. *J. Spacecr.* **1966**, *3*, 182–186. [CrossRef]
7. Williams, J.G. *High-Velocity-Impact Tests Conducted with Polyethylene Terephthalate Projectiles and Flexible Composite Wall Panels*; NASA TN-D-6135; NASA: Washington, DC, USA, March 1971.
8. Hofmann, D.; Christiansen, E.; Lear, D.; Vander Kam, J.; Davis, B. *MSR MMG Debris Cloud HVI Study*, Version 2; NASA Johnson Space Center: Houston, TX, USA, 8 August 2021.
9. Sarli, B.V.; Christiansen, E.; Hofmann, D.; Lear, D.; Vander Kam, J.; Davis, B. *MSR-EEV MMG HVI Test Program*, Version 10; NASA Johnson Space Center: Houston, TX, USA, 17 October 2021.
10. Sarli, B.V.; Christiansen, E.; Hofmann, D.; Lear, D.; Squire, M.; Davis, B. *Enhanced Whipple Shield HVI Study*, Version 7; NASA Johnson Space Center: Houston, TX, USA, 3 March 2022.
11. Sarli, B.V.; Christiansen, E.; Needels, N.; Lear, D.; Stein, R.; Davis, B. *MSR-EEV MMG Shield Evaluation HVI Test Program*, Version 14; NASA Johnson Space Center: Houston, TX, USA, 28 September 2022.
12. Williamsen, J.; Stellingwerf, R.; Evans, S.; Squire, M. Prediction and Enhancement of Thermal Protection Systems from Meteoroid Damage Using a Smooth Particle Hydrodynamic Code. In Proceedings of the 2022 Hypervelocity Impact Symposium, Alexandria, VA, USA, 18–22 September 2022.
13. Rodrigues, R.; Figueiredo, L.; Diogo, H.; Bordado, J. Mechanical Behavior of PET Fibers and Textiles for Stent-Grafts Using Video Extensometry and Image Analysis. *Sci. Technol. Mater.* **2018**, *30*, 23–33. [CrossRef]
14. Rahman, M.; Ayatullah Hosne Asif, A.K.M.; Sarker, P.; Sarker, B. Improvement of Tensile Strength of Viscose Woven Fabric by Applying Chemical Finishes. *Manuf. Sci. Technol.* **2019**, *6*, 23–30. [CrossRef]
15. Schonberg, W.P.; Compton, L.E. Application of the NASA/JSC Whipple Shield Ballistic Limit Equations to Dual-Wall Targets under Hypervelocity Impact. *Int. J. Impact Eng.* **2008**, *35*, 1792–1798. [CrossRef]

Disclaimer/Publisher's Note: The statements, opinions and data contained in all publications are solely those of the individual author(s) and contributor(s) and not of MDPI and/or the editor(s). MDPI and/or the editor(s) disclaim responsibility for any injury to people or property resulting from any ideas, methods, instructions or products referred to in the content.

Article

An Improved Range-Searching Initial Orbit-Determination Method and Correlation of Optical Observations for Space Debris

Xiangxu Lei [1,2], Shengfu Xia [3], Hongkang Liu [3], Xiaozhen Wang [4], Zhenwei Li [5], Baomin Han [1,*], Jizhang Sang [3], You Zhao [2] and Hao Luo [6]

1. School of Civil Engineering and Geomatics, Shandong University of Technology, Zibo 255000, China; xxlei@whu.edu.cn
2. National Astronomical Observatories, Chinese Academy of Sciences, Beijing 100101, China; youzhao@nao.cas.cn
3. School of Geodesy and Geomatics, Wuhan University, Wuhan 430079, China; jzhsang@sgg.whu.edu.cn (J.S.)
4. Xi'an Satellite Control Center, Xi'an 710043, China
5. Changchun Observatory of National Astronomical Observatory, Chinese Academy of Sciences, Changchun 130117, China
6. Shanghai Astronomical Observatory, Chinese Academy of Sciences, Shanghai 200030, China
* Correspondence: hanbm@sdut.edu.cn

Citation: Lei, X.; Xia, S.; Liu, H.; Wang, X.; Li, Z.; Han, B.; Sang, J.; Zhao, Y.; Luo, H. An Improved Range-Searching Initial Orbit-Determination Method and Correlation of Optical Observations for Space Debris. *Appl. Sci.* **2023**, *13*, 13224. https://doi.org/10.3390/app132413224

Academic Editors: Lorenzo Olivieri, Kanjuro Makihara and Leonardo Barilaro

Received: 14 February 2023
Revised: 7 October 2023
Accepted: 6 December 2023
Published: 13 December 2023

Copyright: © 2023 by the authors. Licensee MDPI, Basel, Switzerland. This article is an open access article distributed under the terms and conditions of the Creative Commons Attribution (CC BY) license (https://creativecommons.org/licenses/by/4.0/).

Abstract: The Changchun Observatory of the National Astronomical Observatories, Chinese Academy of Sciences, and the Shanghai Astronomical Observatory are used to generate very short arc (VSA) angle observations of objects in low Earth orbit (LEO) and geostationary orbit (GEO) with their ground-based electrical–optical telescope arrays (EA), the Changchun EA and SAO FocusGEO, respectively. These observations are used in this paper. The range-searching (RS) algorithm for initial orbit determination (IOD) is improved through the multiple combinations of observations and the dynamic range-searching step length. Two different computation modes (the normal mode and the refining mode) of the IOD computation process are proposed. The geometrical method for the association is used. The IOD and association methods are extended to the real optical observations for both LEO and GEO objects. The results show that the IOD success rate of arcs from the LEO objects is about 91%, the error of the semimajor axis (SMA) of the initial orbital elements is less than 50 km, and the correlation accuracy rate is about 89%. The IOD success rate of arcs from the GEO objects is higher than 88%, and the correlation accuracy rate is greater than 87%. The recent COSMOS 1408 antisatellite test (ASAT) generated a large amount of debris. The algorithm of this paper and the observations of Changchun EA are used to initially identify new debris, possibly from the ASAT through initial orbit determination and track association. Finally, 64 suspected new pieces of debris can be found. The results show the effectiveness of the IOD and the correlation algorithm, as well as the potential application of the optical–electrical array in studying space events.

Keywords: space debris; optical telescope; initial orbit determination; track association; orbit determination; antisatellite

1. Introduction

Space debris is defined as man-made objects in space that have lost their function, including parts of failed satellites and spacecraft. Space debris, also known as "space junk", remains above the Earth's atmosphere for many years until it decays, deorbits, disintegrates, or collides with other objects to create new objects [1–3].

Weather forecasting, communications, GPS, and other important space-based everyday life applications are threatened by the increasing volume of orbital debris, all of which depend on a stable space environment [4]. In fact, space debris is already having a real impact on the space environment. As an example, the (International Space Station) ISS has performed 27 collision-avoidance missions since 1999, and, as of July 2021, five of them were to avoid debris from the Cosmos-2251 and Iridium 33 collisions and one to avoid debris from Fengyun-1 [5].

On 15 November 2021, Russia conducted an ASAT test that successfully destroyed a LEO satellite, COSMOS 1408, NORAD ID 13552, weighing about 2.2 tons, with an orbital altitude of 800 km and an inclination of 82.5°. It is estimated that the ASAT event will produce more than 1400 pieces of debris larger than 10 cm in size and more than 70,000 pieces of debris larger than 1 cm in size (from WeChat: Voice of the Chinese Academy of Sciences). As of 10 December 2021, the U.S. has cataloged over 330 pieces of debris from this event in the publicly available TLE catalog, and the number is still growing. It is foreseeable that this ASAT test will aggravate the severe LEO space environment.

There are three methods for observing debris, namely radar, optical observation, and laser ranging. Among these, laser-ranging observation has the highest accuracy [6,7] and is one to two orders of magnitude more precise than microwave radar and optical telescope observations [8]. Optical observation is an important method, and there are a growing number of core key technologies for optical detection being developed and moving from theory to engineering applications, such as super-resolution imaging, polarization spectral detection, and the integration of measurement and detection passes. These advancements provide more efficient and accurate means for detecting space debris [9,10]. Small-sized and medium-high orbit space objects are mainly detected by optical systems, and increasing the telescope aperture can improve the capability to detect faint and weak space objects [8]. The number of space objects since 1957 is shown in Figure 1.

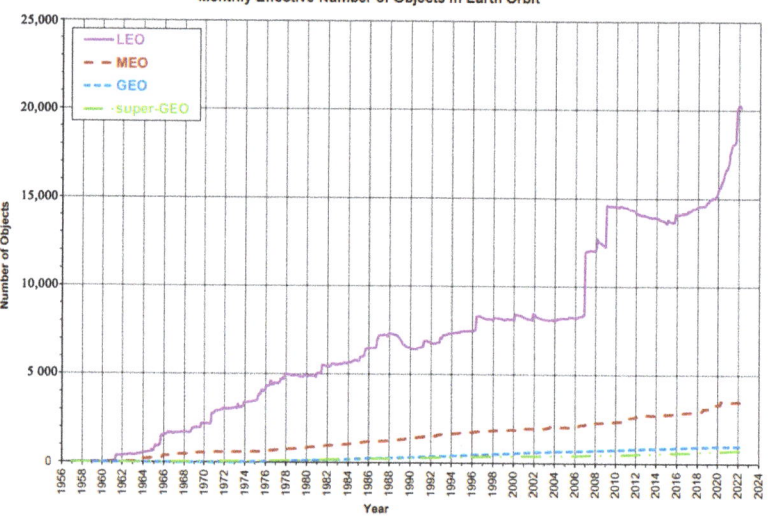

Figure 1. Number of space objects since 1957 [11].

Ground-based optical telescopes are important space surveillance equipment, and reference [12] simulated the capability of multioptical equipment to observe space debris and found that a 15 cm aperture telescope is more valuable for maintaining the space debris catalog. For observations of unknown objects, if these data are used to try to catalog a new object and expand the catalog library, it is common practice to perform IOD and then orbit correlation to achieve the correlated orbit of multiple arcs and finally catalog a new space object. The key techniques include IOD and correlation, and a suitable IOD algorithm can provide the orbit parameters needed for correlation. For the VSA observations, IOD and their correlation are the key techniques for cataloging space debris.

IOD is the rapid calculation of the initial orbit of a space object—a "rough" orbit—using a relatively simple dynamical model (usually a two-body model), without any initial information, using observations of the short-arc segment. There are various methods for IOD, including traditional algorithms such as the Gauss, Laplace, and Gooding methods, as well as modern algorithms such as genetic algorithms. The existing literature on IOD uses a long arc, generally, a few minutes to several minutes; Ref. [13] found that, when the arc length is longer than 400 s, the IOD results are better. With the increase of the arc length, the IOD element errors decrease. But when a certain degree is reached, the error increases and eventually stabilizes. However, the arc length of objects in LEO in real scenarios is often less than 60 s, or even only a dozen seconds, so there are still technical bottlenecks in the IOD with the VSA observations.

There are several algorithms for the initial orbit-correlation problem, and [14] introduced a geometric algorithm for initial orbit correlation for LEO object angle observation data, which were tested using real data, and the results showed that the correct rate of initial orbit correlation for the same object was higher than 80%. Ref. [15] proposed a new method to locate space objects using noncoherent covisual observation techniques, which were validated using optical data from the Changchun, Delingha, and Xuyi stations in the Space Object and Debris Observation Network of the Chinese Academy of Sciences and debris laser-ranging (DLR) data from Changchun station. The results showed that the SMA error of the fixing accuracy was about 1 km. Ref. [16] proposed a new algorithm that does not require initialization and finds a set of observations with minimum residuals using the method of the edge value problem, which can provide complete state and covariance information. Ref. [17] analyzed the influence of the observation geometry on the short-arc angles only IOD and concluded that the optical sensor has the worst surveillance capability for space debris in the same orbital plane. Ref. [18] analyzed the application of algebraic geometry in IOD problems and tested its performance with various scenarios of observations.

The detection of space debris clouds from space events is very significant. With the construction of optical telescopes, more and more optical telescopes are coming into use. This provides a good opportunity to detect events with ground-based telescopes. It is a good idea, but it is difficult to realize and there are few related public studies now. The difficulty lies in obtaining the exact orbit of space debris. To get the exact orbit elements of space debris, the IOD with one short arc and the correlation between the IOD elements are the basis for the detection of space debris.

In this paper, the very short arc IOD and the orbit-correlation study are carried out based on the ground-based optical array of Changchun LEO EA and SAO FoucusGEO. The paper is organized as follows. The first part introduces the background of the paper research. The second part introduces the observation equipment, data, and methods. The third part presents the data-processing results. Finally, the conclusion and outlook are presented.

2. Observations and Methods

2.1. Angle Observations

Usually, an observing device can only track a single object at the same time, and this mode of operation limits the ability to catalog space objects. So, a large field-of-view (FOV) multiobject optical telescope array observation mode was created. The equipment used in this paper is described below (shown in Figure 2 and Table 1).

Figure 2. Physical pictures of Changchun LEO EA ((**a**,**b**), [14]) and Changchun GEO EA and Focus-GEO (**c**).

Table 1. Parameters of telescopes.

Name	Changchun GEO Telescopes	Changchun LEO Telescopes	FocusGEO
Number of telescopes	4	8	3
Telescope diameter	280 mm	150 mm	180 mm
Focus length	324 mm	150 mm	220 mm
CCD	4096 × 4096	3056 × 3056	1528 pixel × 1528 pixel
FOV	6.5° × 6.5°	14.1° × 14.1°	9.5° × 9.5°
Pixel scale of CCD	5.7″/pixel	16.6″/pixel	22.4″/pixel
Angle observation noise	2.4″	5.9″	3″
Detectability	16.5 mag	10.5 mag	15 mag

The Changchun Observatory set up an electrical–optical array (EA) telescope for observing LEO objects and GEO objects at the astronomical observation base in Jilin Province in 2017. The "mini" EA for observing LEO objects consists of eight small telescopes, each with an aperture of 15 cm and a focal length of 15 cm. The system also includes 8 cameras with 3 × 3 K resolution, 8 image processing computers, 1 GPS clock, an electronic control system, an image acquisition and processing system, etc. The monitored sky area is up to 1590 square degrees, mainly observing space debris with elevation angles from 18 to 32 degrees in the sky area with diameters ranging from 0.5 to 1 m.

The optical telescopes used for observing space debris in GEO orbit are part of an all-sky rotatable array, which is designed for observing space debris. The array consists of four 280 mm optical telescopes and two T-frame structure equatorial instruments, with an average error of 0.9″ in the right ascension direction and 1″ in the declination direction in terms of tracking measurement accuracy. With the same design parameters, the four telescopes can significantly expand the sky area, covering 160 square degrees, and detect objects with brightnesses between 16 and 20 magnitude.

At the end of 2017, SAO developed the FocusGEO, an EA with a large FOV dedicated to detecting GEO objects, which consists of three 0.18/0.22 m refracting optical cylinders, forming a large rectangular FOV of 9.5° × 28.5°, and can scan the GEO belt of about 3200 square degrees above the observatory in 15 min. The angle observations by the FocusGEO telescope at the Lijiang Observatory of the Yunnan Astronomical Observatory show that the telescope can observe nearly 380 GEO space objects over the station during non-full-moon clear nights. The observational capabilities of FocusGEO surpass 90% for the objects in GEO orbit that are cataloged in the NORAD catalog. With an exposure of 5 s, the telescope can detect GEO objects of 15 magnitude with a precision of about 3″ in azimuth and pitch.

2.2. The Improvement of the RS IOD Method and the Association Method

2.2.1. The Improvement of the RS IOD Method

The IOD orbit is determined using the range-searching (RS) method, and the basic process of the range-searching method is as follows [14]. Suppose that the angular observations of a space object are obtained at moments $t_1, t_2, \cdots, t_n$, right ascension $RA_1, RA_2, \ldots, RA_n$ and decimal declination $Dec_1, Dec_2, \cdots, Dec_n$, respectively. The unit vector L_i ($i = 1, 2, \cdots, n$) of the direction from the station to the space object (i.e., the telescope line of sight direction) can be obtained by the following equation,

$$\begin{cases} Lx_i = \cos(Dec_i)\cos(RA_i) \\ Ly_i = \cos(Dec_i)\sin(RA_i) \\ Lz_i = \sin(Dec_i) \end{cases} \quad (1)$$

Lx_i, Ly_i, Lz_i are three components of the coordinates of three directions.

With ρ_1 and ρ_n, denoting the observed distances at moments r_1 and r_n, respectively, and R_i representing the position vector of the observing station, if we know ρ_1 and ρ_n, we can get two position vectors r_1 and r_n based on $r_i = R_i + \rho_i L_i$ ($i = 1, n$), so that the purely angular IOD problem is converted into an orbit-determination problem based on two position vectors, which is the so-called IOD Lambert problem (the orbital parameters are calculated using the position vectors of the two moments). The Lambert problem is well known in dynamical astronomy, celestial mechanics, and astrodynamics communities for objects governed by Keplerian dynamics [19]. The classical methods can be used to solve the standard Lambert problem given two positions, such as the Gibbs method or the Herrick–Gibbs method [20,21]. If the values of ρ_1 and ρ_n are set to a fixed step-size sequence within a specified range, each combination of observations can be used to compute a set of orbital parameters. Subsequently, the calculated angular observations at other moments are compared with the actual observations and judged based on the residuals. This helps to eliminate the erroneous track determination results and narrows down the options to the most likely ones. Finally, the optimal solution is selected based on constraints, such as the semimajor axis (SMA) and eccentricity, within the bounds set by the limitations of the space environment. The original specific implementation process can be referred to in [22].

To improve the success rate and accuracy of IOD, the RS method is modified through the following steps.

Firstly, as many combinations of observations as possible are made over the longest possible epoch range. Since observation errors vary across different epochs, the observations at different epochs can lead to different combinations and result in different potential orbit elements. The greater the number of combinations, the higher the success rate. The iteration steps are outlined in the following pseudocode (Algorithm 1):

Algorithm 1: The iteration steps of IOD computation.

For $i = 1, 2, 3, n/2$, $i = i + step$ (default 1), n is the epoch number.
    For $j = 2/n, n$, $j = j + step$ (default 1)
    Determine the used observations (RA_i, Dec_i) and $\left(RA_j, Dec_j\right)$. Observations at the other
        epochs of the current arc are used as a discriminator.
    Solving the Lambert problem to get a set of orbit elements. If the residuals are less than the
        threshold preset, that means a possible solution. Then, a set of elements is added to the result sum.
    End
End

Secondly, to improve the accuracy of the IOD elements, the key is the accuracy of the distance. The distance is estimated and used to compute the elements again; then, a new set of orbital elements is achieved. As a possible result, it may eventually replace the previous one. A simulation test is implemented to show the influence of the observed distance on the accuracy of IOD solution elements. So, if the distance is error free or with very limited errors, the IOD elements' accuracy is limited.

The previous method was stable, but the step length had an impact on the orbit solutions. To enhance the accuracy, the new method features a dynamic step-length adjustment based on the varying thresholds of the observation residuals and the corresponding step length. This ensures that the solution is optimized, and the accuracy is improved. The computation process is divided into two modes, the normal mode and the refining mode. There are two sets of thresholds for the two modes, respectively.

The used threshold values $thres_{used}$ of observation residual res_i are set as follows,

$$thres_{used} = \begin{cases} \text{next step,} & (\text{if } res_i > thres_{normal}) \\ thres_{normal}, & (\text{if } res_i < thres_{normal} \text{ and } res_i > thres_{refine}) \\ thres_{refine}, & (\text{if } res_i < thres_{refine}) \end{cases} \quad (2)$$

$thres_{normal}$ is the residual threshold at the normal mode, and $thres_{refine}$ is the residual threshold at the refining mode. The $thres_{refine}$ is set to be $thres_{refine} = coefficient \times thres_{normal}$; the coefficient is a fraction less than one and is usually set to 0.5.

The used step length $step_{used}$ are set as follows,

$$step_{used} = \begin{cases} step_{normal}, & (\text{if } res_i > obs_{normal}) \\ step_{normal}, & (\text{if } res_i < obs_{normal} \text{ and } res_i > obs_{refine}) \\ step_{refine}, & (\text{if } res_i < obs_{refine}) \end{cases} \quad (3)$$

$step_{normal}$ is the step length at normal mode and $step_{refine}$, means the step length at refining mode.

We can determine the values of the used threshold and step by following the steps. First, set the value of two normal thresholds based on experience, and set $thres_{used} = thres_{normal}$ and $obs_{used} = obs_{nomral}$. Second, for the ith combination of observations, compute orbit element sets, and the observed residuals res_i are calculated by orbital propagation. Third, the step length and the threshold used at $(i+1)^{th}$ depend on the res_i and are computed according to the above two Equations (2) and (3).

The aforementioned thresholds and steps in both the normal and refining modes are critical to the search success rate, IOD precision, and effectiveness. However, these values are not usually fixed and must be determined through extensive experiments for different scenarios. For example, they can be obtained from ground-based and space-based surveillance simulations and data-processing experiences.

2.2.2. The Association Method

The geometrical method is used for the correlation of the initial orbital elements, the core step of this correlation algorithm, and the specific implementation process can be referred to [14]. The association technique is founded on the fact that the error of SMA accumulates to produce an along-track bias at the midpoint between two propagated positions, which are estimated from two IOD solutions. By iteratively modifying the SMAs of these solutions until the errors in the SMAs are of equal magnitude but opposite signs, the along-track bias at the midpoint approaches zero if the two orbit-determination solutions pertain to the same object.

The core ideas of the initial track-association method used in this paper are as follows.

(1) Determine the two sets of initial orbit elements for conducting the association;
(2) The two tracks are propagated to the intermediate moments of the two by using the analytical method of orbit propagation. This allows for the correlation of the two tracks;
(3) After propagation, the differences between the two initial tracks are calculated in the along-track, cross-track, and radial directions (ACR). The semimajor axis (SMA) of the two initial tracks is then adjusted based on the differences in the along-track direction. The tracks are repropagated, and the differences in the ACR directions of the two initial tracks are recalculated;
(4) The final ACR difference is judged after applying multiple corrections in succession, and, if it is less than the preset threshold, the two tracks are considered to be from the same object. Otherwise, they are from different objects.

3. Results

3.1. IOD of Arcs from LEO Objects

First, the ground-based optical–electrical array observations from Changchun Observatory were used for TLE matching to identify the known objects. TLE data are public. The results are shown in Table 2, where about 85% of the arc segments were from known objects. The average arc length of the observed data is about 39 s.

Table 2. Number of the identified space objects and the arcs.

Date	Number of the Arcs	Number of Arcs from Known Objects	Rate	Number of Known Objects
24 August 2017	4100	3458	84.34%	1299
25 August 2017	1626	1396	85.85%	594
26 August 2017	4894	4163	85.06%	1587

The errors are estimated from the TLE of the known debris. TLE can be regarded as a reference or true orbit for the IOD elements, and it is more precise than IOD elements. So, the IOD errors can be calculated by comparing the TLE orbit and the IOD elements at a common epoch for space debris. Then, the IOD elements are estimated, and the initial orbit results of the LEO objects are shown in Table 3, which contains the initial orbit-determination success rate and the SMA errors. A set of 7488 orbits was used, and the success rate of initial orbit determination was about 95.6%, higher than the previous success rate in [14]. Finally, the initial orbit correlation was performed, and the correct rate of initial orbit correlation was about 89%.

Table 3. IOD results of space debris in LEO.

Errors of SMA	Number	Rate
<100 km	6752	90.17%
<50 km	6099	81.45%
<30 km	5150	68.78%
<20 km	4088	54.59%
<10 km	2294	30.64%
<5 km	1143	15.26%

3.2. IOD of Arcs from GEO Objects

3.2.1. Changchun GEO EA

In this paper, the test is carried out using observation data from 6–10 February 2021. The arc length distribution of the observation data is shown in Figure 3. The arcs are matched with the TLE to identify the cataloged objects. The TLE data were downloaded from www.space-track.org, (accessed on 7 February 2023). The matching results are shown in Table 4, with approximately 74.3% of the arc segments being known objects. Based on the data and the results of the TLE matching, the number of observed arcs for each object can be analyzed, and the statistical results are shown in Figure 3, where it can be found that the number of observed arcs for most objects is no more than 100.

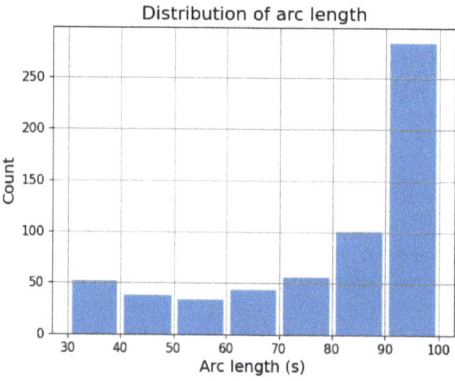

Figure 3. Distribution of arc length.

Table 4. The matching result between the arc observation and TLE.

Date	Number of Arcs	Number of Matched Arcs	Rate (%)
6 February 2021	25,714	17,369	67.50
7 February 2021	15,513	11,302	72.86
8 February 2021	14,250	10,960	76.91
9 February 2021	11,976	9355	78.11
10 February 2021	16,709	12,734	76.20
Mean	16,832	12,344	74.32

Some of the known observation arcs were selected for IOD, and the relationship between the IOD success rate and arc length was initially analyzed. The arc length of GEO object data was longer than 30 s, and the average arc length was about 77.4 s, with a total number of 1602 successful IODs and a success rate of 87.84%. For the observations in which the TLE was matched successfully, the TLE was used to evaluate the initial orbit parameter errors, using the SMA and inclination as examples, and the results are shown in Tables 5 and 6. According to Table 5, it can be found that 40% of the total SMA errors are less than 30 km, and about 71% of them are less than 100 km. According to Table 6, it can be found that 82% of the inclination errors are less than 0.1 (the total number of IODs is 539).

Table 5. The statistics for the IOD SMA errors of space objects in GEO.

Errors of SMA	Count	Rate
<100 km	962	70.58%
<50 km	743	54.51%
<30 km	546	40.06%
<10 km	232	17.02%

Table 6. The statistics for the IOD inclination errors of space objects in GEO.

Error (Deg)	Count	Rate
<1	523	97.03%
<0.5	515	95.55%
<0.2	488	90.54%
<0.1	444	82.37%

When the time interval is less than 36 h, the correlation rate of arcs of the same object is about 89%. During the data processing, some problems were found. Further preprocessing was needed, such as the duplication of data between observation files or failure to correlate consecutively observed arc segments of the same object, splitting into two or more files. The statistics revealed that many of the observation arcs for a given object had small intervals between them or even overlapped.

3.2.2. FocusGEO

The data were observed on 20, 22, and 23 October 2019, and 5002, 5093, and 4086 arcs were obtained from each day. Then the arcs are tried to match with the TLE data and the IOD elements are computed. The results are shown in Table 7 and Figure 4.

Table 7. The statistical errors of IOD SMA, eccentricity, and inclination of observations observed by FocusGEO.

SMA Error (km)	Rate (%)	Eccentricity Errors	Rate (%)	Inclination Errors (Deg)	Rate (%)
<5	4.12	<0.0001	21.70	<0.01	11.26
<10	8.24	<0.0005	47.80	<0.03	31.32
<30	21.15	<0.001	53.02	<0.05	40.66
<50	33.24	<0.005	82.42	<0.1	64.29
<100	55.49	<0.01	97.53	<0.2	83.79

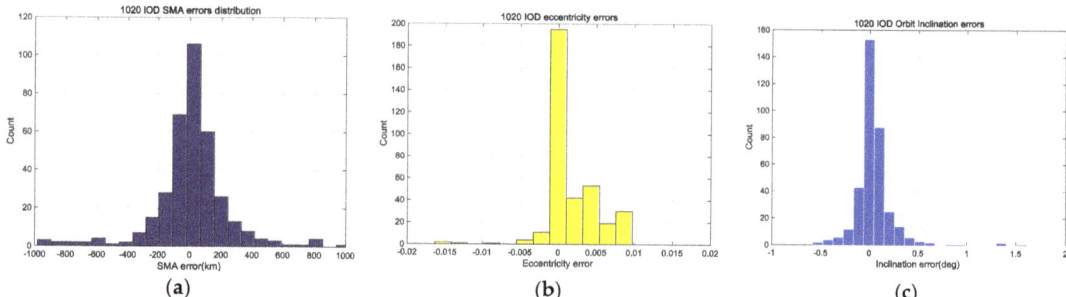

Figure 4. The error distribution of IOD SMA, eccentricity, and inclination of observations observed by FocusGEO. (**a**) SMA; (**b**) eccentricity; (**c**) inclination.

Selected for calculation were 881 arcs containing 9 or 10 data points from the observation file of 20 October 2019, and the arc lengths were all about 53 s. There were 868 arcs with successful orbit computations, and the success rate was 98.52%. The TLE data were used to evaluate the initial orbit errors, and the results are shown in Table 7. The corresponding error distributions of SMA, eccentricity, and inclination errors are shown in Figure 4. It can be found that over 55% of the IOD SMA errors are less than 100 km for GEO objects and about 53% of all the IOD eccentricity errors are smaller than 0.001.

As to the IOD success rate, the rate of Changchun GEO is higher than that of FocusGEO. The reason should be that the average arc length is different. The average length of the former is about 77 s but the average length of the latter is about only 53 s. This is also the reason for the worse result of the IOD elements solution.

3.3. IOD of Observations of Space Debris Related to the COSMOS 1408 Satellite

3.3.1. Background

Space events are happening with increasing frequency, such as the breakup events of the YunHai 1-02(2019-063A, NORADID 44547), the NOAA 17(2002-032A, NORAID 27453), the ASAT 1408 in 2021, and the American GEO satellite Galaxy 11 (1999-071A, NORADID 26038) in 2022 (from ISON). The antisatellite incident had the biggest impact. On 15 November 2021, at 10:47 p.m. Beijing time, Russia conducted an antisatellite test (ASAT) that destroyed the COSMOS 1408 satellite. The satellite, a former Soviet satellite, was launched in 1982 with NORAD ID 13552. The orbit of Cosmos 1408 prior to the ASAT test had an altitude of 490 × 465 km. It is estimated that more than 1500 trackable pieces of space debris, as well as countless smaller pieces, were generated from the ASAT event. This posed a threat to the security of space assets of countries around the world. The debris number evolution and the occurred operations are shown in Figure 5.

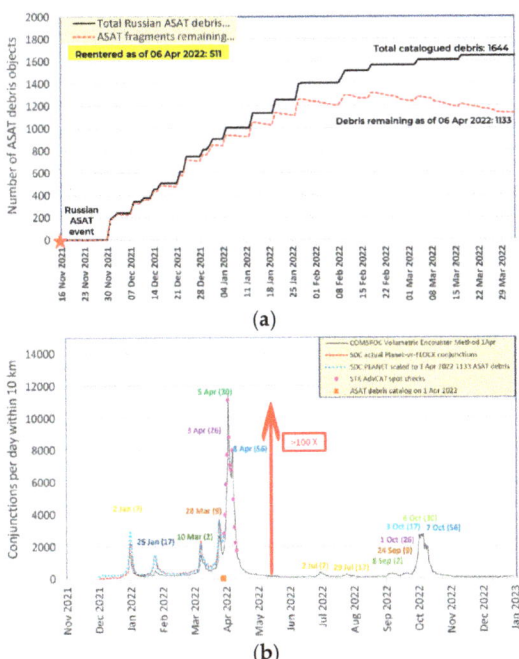

Figure 5. The Russian ASAT conjunction squall predications vs. SDC operational results (from COMSPOC). (**a**) The COSMOS 1408 debris fragment tracking and decay evolution. (**b**) Daily Encounters between Planet's 233 spacecraft and ASAT debris.

As of 26 May 2022, 900 pieces of debris from this antisatellite event, with a cataloged target of 1740, have fallen, and 818 will fall in the next 3 years (SATEVO). The estimated in-orbit times of debris of different sizes generated by antisatellite events in NASA's Space Debris Quarterly Report are shown in Figure 6. Based on the TLE from http://celestrak.org, (accessed on 7 February 2023), the number of pieces of space debris created in this ASAT event is 260 on 12 February 2023.

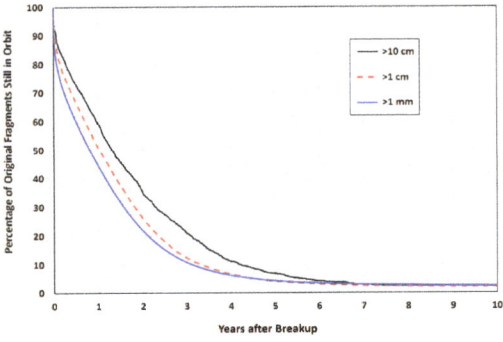

Figure 6. The predicted orbital decay of the Cosmos 1408 fragment cloud. The three curves are, from top to bottom, ≥ 10 cm fragments, ≥ 1 cm fragments, and ≥ 1 cm fragments, respectively [11].

3.3.2. IOD of Unmatched Arcs of Space Objects

First, the observation data is tried to match with TLE to judge whether the observation arc is from a new object or a cataloged object. The SMA, eccentricity, and inclination of the COSMOS 1408 satellite orbit are 6,862,203.9 m, 0.00285, and 82.57°, respectively.

For the part of the observation data of the past two days that cannot be successfully matched with the TLE, the IOD is carried out. Due to the approaching full moon, object imaging was affected to a certain extent, and the amount of data on 16 November was relatively small. Due to the weather, there were no observations on 17 November. In the initial orbit results, the orbit parameters whose SMA is between 6800–6950 km and the inclination angles between 75°–90° are screened out. Counting the results of three days, respectively, it was found that only the COSMOS 1408 satellites were in this interval on 14 November, and 64 and 9 initial orbits were in this interval on 15 November and 16 November, respectively. The orbit parameter distribution is shown in Figure 7.

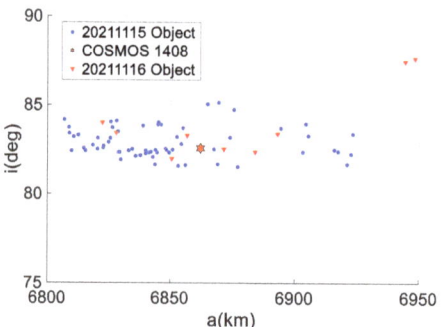

Figure 7. IOD parameters of observations on 15 November 2021 and 16 November 2021.

The number of objects observed near the orbital altitude and inclination of the COSMOS 1408 satellite is analyzed. The observation arcs that have been successfully matched with the TLE object are excluded. According to Figure 7, before the event, there was no object in the related sky area on the night of 14 November. Following the incident, 64 objects appeared in the airspace on the night of 15 November, and 9 objects appeared on the night of 16 November. Therefore, the observations on the 15th–16th are likely to be debris generated by the COSMOS 1408 event. The 64 initial orbits were correlated. One pair was successfully correlated, and the orbit-determination results of the two arc segments showed that the inclination angle was about 82.33°.

Further analysis of the initial orbit results on 16 November shows that there are a large number of objects with an inclination angle of 53° and an SMA between 6880 km and 6950 km (Figure 8). It is noted that 53 Star-link satellites were launched by SpaceX at 12:40 (UTC) on 12 November 2021 before the ASAT event. But, the TLE of these satellites was not public at that time, so they failed to match with TLEs.

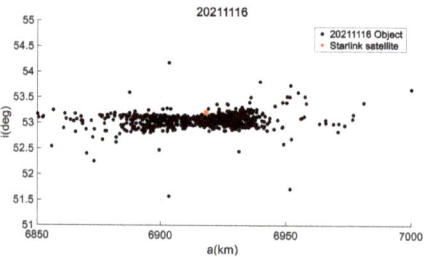

Figure 8. IOD parameters of observations for Star-link satellites on 16 November 2021.

4. Discussion

Optical telescopes are important devices for monitoring space objects. Two key techniques for cataloging new objects using angle observation data are IOD and IOD correlation. In response to these two problems, this paper uses the improved range-searching (RS) algorithm and geometrical method to determine the initial orbital elements and conduct track correlation. The RS method is improved through the new computation modes with different threshold constraints proposed in this paper. They are so-called the "normal mode" and "refining mode". Accordingly, to improve the accuracy of the method, the corresponding step length for range searching is designed also. Then, the RS method and the geometrical method are extended to the real observations at multistations for both LEO and GEO objects. The observations used in this study were obtained from the ground-based optical–electrical array (EA) of the Changchun Observatory of the Chinese Academy of Sciences and the Shanghai Astronomical Observatory. Observations of both LEO and GEO objects observed by these ground-based VSAs were utilized. The results show that the success rate of the IOD of the LEO object is about 91%, 81% of the initial orbit SMA errors are less than 50 km, and the correlation accuracy rate is about 89%; the success rate of the GEO object IOD is higher than 88%, 54% of the errors of the SMA of the initial orbit is less than 50 km, and the correlation accuracy rate is greater than 87%. The measured data-processing results show that the IOD and correlation algorithm in this paper is suitable for high and low-orbit space-object observation data.

Further, this paper briefly introduces Russia's ASAT event that occurred on 15 November 2021, and analyzes the resulting debris orbit information. The preliminary results of identifying the new debris generated by COSMOS 1408 based on ground-based observation data in Changchun show that the correlation orbit-determination algorithm in this paper can be used to quickly discover new objects. For space objects in LEO, optical data can be used as a useful supplement to radar data, and the experimental results further demonstrate the potential of EA. Limited by certain conditions, this paper only conducts a preliminary analysis of COSMOS 1408-related fragments and more observation data will be used for further research in the future.

Author Contributions: Conceptualization, X.L., J.S. and X.W.; origination of the idea of this paper; methodology, X.L., S.X., B.H. and Y.Z., the improved method is discussed by these authors and implemented by X.L. and S.X.; software, S.X. and X.L., they develop the software; validation, H.L. (Hongkang Liu); formal analysis, Y.Z. and X.W., they analyze the original manuscript and proposed valued suggestions for the method; investigation, X.L. and Y.Z., they investigate the IOD methods and discussed the new method; resources, B.H., Z.L., Y.Z. and H.L. (Hao Luo), they coordinate and provide the angle observations from several stations used in this paper; data curation, Z.L., Y.Z. and H.L. (Hao Luo), they provide the angle observations used in this paper; writing—original draft preparation, X.L.; writing—review and editing, X.L., S.X. and H.L. (Hongkang Liu), modify and polish the manuscript; visualization, X.L. and H.L. (Hongkang Liu); the figure creation; supervision, X.L. and S.X.; project administration, X.L.; funding acquisition, J.S. and X.L. All authors have read and agreed to the published version of the manuscript.

Funding: This research was funded by Shandong Provincial Natural Science Foundation General Project, grant number ZR2023MD098, the Key Laboratory of Geospace Environment and Geodesy, Ministry of Education, Wuhan University, grant number 21-01-02, Scientific Innovation Project for Young Scientists in Shandong Provincial Universities, grant number 2022KJ224, Chongqing Municipal Natural Science Foundation of General Program, grant number CSTB2022NSCQ-MSX1093, Shanghai Sailing Program, grant number 21YF1455200 and the Youth Innovation Promotion Association CAS, grant number 2023273.

Institutional Review Board Statement: Not applicable.

Data Availability Statement: The data presented in this study are available on request from the corresponding author. The data are not publicly available due to the following reason relevant state regulations of the data provider.

Conflicts of Interest: The authors declare no conflict of interest.

References

1. Li, M.; Gong, Z.; Liu, G.Q. Frontier technology and system development of space debris surveillance and active removal. *Chin. Sci. Bull.* **2018**, *63*, 2570–2591. [CrossRef]
2. Gong, Z.; Song, G.; Li, M.; Wang, J.; Huang, Y. Long-term Sustainability of Space Activities: From Space Traffic Management to Space Environment Governance—Review on the 683rd Xiangshan Science Conference. *Space Debris Res.* **2021**, *21*, 5–12. (In Chinese)
3. Tang, J.S.; Cheng, H. The origin, status and future of space debris. *Physics* **2021**, *50*, 317–323.
4. NASA. *NASA's Effort to Mitigate the Risks Posed by Orbital Debris*; Report No. IG-21-011; 27 January 2021; NASA: Washington, DC, USA, 2021.
5. Anz-Meador, P.D. *Orbital Debris Quarterly News*; National Aeronautics and Space Administration: Washington, DC, USA, 2020; Volume 24.
6. Li, G.; Liu, J.; Cheng, H. Space Debris Laser Ranging Technology and Applications. *Space Debris Res.* **2020**, *20*, 40–48. (In Chinese)
7. Long, M.; Deng, H.; Zhang, H.; Wu, Z.; Zhang, Z.; Chen, M. Development of Multiple Pulse Picosecond Laser with 1 kHz Repetition Rate and Its Application in Space Debris Laser Ranging. *Acta Optica Sin.* **2021**, *483*, 155–162. (In Chinese)
8. Zhang, D. Dim Space Target Detection Technology Research Based on Ground-Based Telescope. Ph.D. Thesis, University of Chinese Academy of Sciences, Beijing, China, 2020.
9. Fu, Q.; Shi, H.; Wang, C.; Liu, Z.; Li, Y.; Jiang, H.-L. Research on New Technology of Photoelectric Detection for Space-Based Space Debris. *Space Debris Res.* **2020**, *20*, 49–55. (In Chinese)
10. Sun, R.; Zhao, C. Optical Survey Technique for Space Debris in GEO. *Prog. Astron.* **2012**, *30*, 394–410.
11. Cowardin, H. *Orbital Debris Quarterly News*; National Aeronautics and Space Administration: Washington, DC, USA, 2022; Volume 26.
12. Hu, J.; Hu, S.; Liu, J.; Chen, X.; Du, J. Simulation Analysis of Space Debris Observation Capability of Multi-Optoelectronic Equipment. *Acta Opt. Sinic.* **2020**, *468*, 29–35. (In Chinese)
13. Liu, L. Study on the Initial Orbit Determination of Space Targets with Space-Based Surveillance. Ph.D. Thesis, Graduate School of National University of Defense Technology, Changsha, China, 2010.
14. Lei, X.; Li, Z.; Du, J.; Chen, J.; Sang, J.; Liu, C. Identification of uncatalogued LEO space objects by a ground-based EO array. *Adv. Space Res.* **2021**, *67*, 350–359. [CrossRef]
15. Chen, L.; Liu, C.; Li, Z.; Sun, J.; Kang, Z.; Deng, S. Non-Cooperative Common-View Observation of LEO Space Objects and Initial Orbit Determination. *Acta Optica Sin.* **2021**, *496*, 162–167. (In Chinese)
16. Pastor, A.; Sanjurjo-Rivo, M.; Escobar, D. Initial orbit determination methods for track-to-track association. *Adv. Space Res.* **2021**, *68*, 2677–2694. [CrossRef]
17. Feng, Z.; Yan, C.; Qiao, Y.; Xu, A.; Wang, H. Effect of Observation Geometry on Short-Arc Angles-Only Initial Orbit Determination. *Appl. Sci.* **2022**, *12*, 6966. [CrossRef]
18. Mancini, M. An Analysis on the Application of Algebraic Geometry in Initial Orbit Determination Problems. Master's Thesis, Georgia Institute of Technology, Atlanta, GA, USA, 2022.
19. Russell, R.P. On the solution to every Lambert problem. *Celest. Mech. Dyn. Astron.* **2019**, *131*, 50. [CrossRef]
20. Escobal, P.R. *Methods of Orbit Determination*; Wiley: New York, NY, USA, 1965.
21. Vallado, D.A.; McClain, W.D. Fundamentals of Astrodynamics and Applications. In Proceedings of the Tutorial Lectures at the 4th ICATT, Madrid, Spain, 30 April 2010.
22. Zhang, P.; Sang, J.; Pan, T.; Li, H. Initial 0rbit Determination Method Based on Range Searching for LEO Space Debris. *Spacecr. Eng.* **2017**, *26*, 22–28. (In Chinese)

Disclaimer/Publisher's Note: The statements, opinions and data contained in all publications are solely those of the individual author(s) and contributor(s) and not of MDPI and/or the editor(s). MDPI and/or the editor(s) disclaim responsibility for any injury to people or property resulting from any ideas, methods, instructions or products referred to in the content.

Article

Research on the Efficient Space Debris Observation Method Based on Optical Satellite Constellations

Gongqiang Li [1,2,3,*], Jing Liu [1,2,3], Hai Jiang [2,3] and Chengzhi Liu [1,2]

1. Changchun Observatory, National Astronomical Observatories, Chinese Academy of Sciences, Changchun 130117, China
2. University of Chinese Academy of Sciences, Beijing 100049, China
3. National Astronomical Observatories, Chinese Academy of Sciences, Beijing 100101, China
* Correspondence: ligongqiang@bao.ac.cn

Abstract: The increasing amount of space debris poses a major threat to the security of space assets. The timely acquisition of space debris orbital data through observations is essential. We established a mathematical model of optical satellite constellations for space debris observation, designed a high-quality constellation configuration, and designed a space debris tracking observation scheduling algorithm. These tools can realize the efficient networking of space debris from a large number of optical satellite observation facilities. We designed a constellation consisting of more than 20 low-Earth orbit (LEO) satellites, mainly dedicated to the observation of LEO space objects. According to the observation scheduling method, the satellite constellation can track and observe more than 93% of the targets every day, increase the frequency of orbital data updates, and provide support for the realization of orbital space debris cataloguing. Designing optical satellite constellations to observe space debris can help realize the advance perception of dangerous collisions, timely detect dangerous space events, make key observations about high-risk targets, greatly reduce the false alarm rate of collisions, and provide observational data support for space collisions.

Keywords: constellation; satellite; observation; space debris; scheduling

Citation: Li, G.; Liu, J.; Jiang, H.; Liu, C. Research on the Efficient Space Debris Observation Method Based on Optical Satellite Constellations. *Appl. Sci.* **2023**, *13*, 4127. https://doi.org/10.3390/app13074127

Academic Editors: Lorenzo Olivieri, Kanjuro Makihara and Leonardo Barilaro

Received: 16 February 2023
Revised: 7 March 2023
Accepted: 20 March 2023
Published: 24 March 2023

Copyright: © 2023 by the authors. Licensee MDPI, Basel, Switzerland. This article is an open access article distributed under the terms and conditions of the Creative Commons Attribution (CC BY) license (https://creativecommons.org/licenses/by/4.0/).

1. Introduction

Space debris refers to the non-functional man-made objects in orbit. Human space activities have a history of more than 60 years, and it is estimated that there are over 30,000 pieces of debris larger than 10 cm, approx. one million pieces of debris larger than 1 cm, and hundreds of millions of millimeter-level space debris. Space debris and spacecraft move around the Earth at a high speed of more than 7.9 km/s, and the relative speed of space debris can reach more than 10 km/s when a collision occurs. The impact of a 1 cm-sized aluminum ball in space will produce a destructive force equivalent to the impact of a car on the highway. The consequences of the impacts of space debris above 1 cm are often devastating, and large-sized space debris above 10 cm will directly lead to spacecraft failure [1–3]. Space collisions have become more frequent in recent years. In 2019, there were more than 300 dangerous impacts between Chinese spacecraft in orbit and space debris. In 2013, the Ecuadorian CubeSats collided with Soviet rocket debris, causing the satellite to fail. On 12 June 2011, the solar cells of the IGSO-2 satellite of China's Beidou navigation constellation lost two circuits, and the conclusion confirmed that the satellite was hit by small debris. On 11 February 2009, the US commercial communications satellite Iridium-33 collided with Russia's abandoned Cosmos-2251 satellite, with a relative speed of 11.6 km per second, generating more than 2000 pieces of space debris, which caused strong repercussions from the international community [4].

With the development of human space activities, the number of space objects is increasing. At the same time, the development of large satellite constellations and the competition for space interests of various countries have made the space situation increasingly complex [5]. This poses a major threat to the security of space activities and space assets. It is necessary to obtain the orbit information of space objects in real time to meet the demand for accurately detecting space events [6]. Monitoring space objects using monitoring equipment and optimizing the scheduling of the monitoring system is an effective way to obtain the space object's status information [7]. By adopting efficient observation scheduling methods, the number of space objects and arcs observed by observation equipment in a certain period can be increased, and the observation efficiency can be improved [8].

Compared to traditional ground-based optical telescopes, space-based optical satellites have unique advantages because they run in orbit and their detection of space objects are not affected by the weather. In addition, the detector is closer to the space object, and an optical telescope with a smaller aperture can be used to observe the low-orbit space target at a shorter distance. Most space-based satellite constellations are deployed in low Earth orbit where space debris is mainly distributed. Therefore, satellite constellations can fully observe the low Earth orbit airspace where space objects are densely distributed, improve the detection quantity and frequency of space objects, make full use of space detection equipment, obtain a large amount of observation data in real time, realize the dynamic catalogue of space debris, provide early warning of dangerous collisions, and maintain the order of space traffic and the safety of space assets [9].

Reasonable monitoring task scheduling is the key for the effective operation of a space-based monitoring network. Especially due to the development of various space-based monitoring equipment and the improvement of the automation of observation equipment, new requirements have been put forward to optimize observation task scheduling. Space object observation scheduling optimization is bound to be closely related to the specific situation of equipment operation and the movement of the space object itself. Special research needs to be conducted on the specific application observation mode and the equipment operation state. This is a very complex combinatorial optimization problem that exhibits multi-time window constraints, multi-resource constraints, and high conflict, and it has become one of the leading issues in the field of space monitoring [10–12].

The current observation scheduling algorithm is not suitable for rapidly moving observation equipment and observation targets, and there is little research on the observation task scheduling of space-based satellite constellations. The research on the observation task scheduling algorithm of space-based optical satellite detecting for space objects is the key for improving the efficiency of space object detection, and it is also one of the important problems that needs to be solved in the field of space object observation [13–15].

We designed satellite constellations with different configurations and created a constellation observation task scheduling algorithm. The main purpose was to track and observe LEO space objects, obtain a large amount of observation data, and update the object orbit information in real time. The goal was to catalog a large number of LEO space objects using a constellation composed of a small number of satellites. The space objects orbit data were used for the simulation and verification, and good results were obtained. With the observation of satellite constellation, we can realize space situation awareness, provide emergency response to space emergencies, reduce collision risks, and maintain the safety of space assets.

2. Constellation Design

2.1. Model Establishment

The constellation spatial coverage model was established by comprehensively considering the point coverage numerical simulation model, the optical sensor coverage calculation, ground shadow model, the visual function, and the constellation comprehensive coverage performance evaluation index [16]. The point coverage numerical simulation technology was used to divide the target airspace into multiple blocks according to specific criteria. According to the constellation coverage of all the blocks, the space and time coverage performance of the constellation to the target airspace was analyzed. In order to avoid the infrared radiation interference of the Earth and its atmosphere, the optical sensors on the satellite could only face the cold background of the universe to detect space objects above the edge of the Earth. The influence of the Earth's shadow on the observation of space debris also needs to be considered. The coverage of the constellation to the target airspace was measured in time and space to evaluate the comprehensive coverage performance of the constellation.

(1) Point coverage numerical simulation model

The numerical simulation technique of the point coverage is also known as the spatial meshing method [17], as shown in Figure 1. The target airspace is divided into grid points according to specific criteria, and the coverage of the target airspace in time and space is analyzed according to the coverage of all the airspace grid points by the constellation.

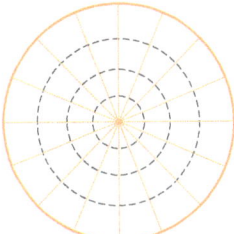

Figure 1. Schematic diagram of the point coverage grid method.

The problem surrounding constellation coverage in general Earth observation is that it mainly considers the two-dimensional coverage of the earth's surface. Due to the need to monitor space debris, two-dimensional grid point sampling needs to be extended to three-dimensional, and low-orbit space-based optical satellite constellations need to consider three-dimensional coverage within a certain altitude range above the Earth's surface where the sampled airspace is roughly a spherical shell. To establish a three-dimensional grid point sampling criterion, the height variables and the sample latitude, longitude, and altitude at the intervals need to be added.

(2) Optical sensor coverage calculation

In order to avoid infrared radiation interference from the Earth and its atmosphere, the optical sensor of the satellite can only be oriented towards the cosmic cold background of space to detect objects above the edge of the Earth [18], as shown in Figure 2.

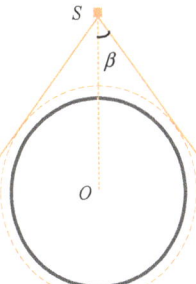

Figure 2. Coverage area of the space-based optical sensor.

In Figure 2, O is the center of the earth and S is the space-borne optical sensor. The solid circle represents the sphere model of the Earth, and the dotted circle represents the edge of the Earth's atmosphere. The sensor covers a sector beyond the tangent line of the satellite to the Earth's atmosphere.

The covering conditions of a space-based optical sensor for specific space debris in a specific altitude are shown in Formula (1).

$$\begin{cases} \varphi > \beta \\ 0 < d_{st} < L_{max} \end{cases} \quad (1)$$

where φ is the angle between the space object and the Earth's center relative to the satellite, β is the angle between the earth's atmosphere boundary and the Earth's center relative to the satellite, d_{st} is the distances between the satellites and the space object, and L_{max} is the maximal detection range of the satellite-borne optical sensor.

(3) Earth shadow model

When space-based optical sensors observe space objects, the space objects need to be illuminated by the Sun. When the space object is in the shadowed area, the optical sensor cannot observe it, so the influence of the shadow on the observation of the space object needs to be considered.

Since the Sun is far from the Earth, the solar beam is regarded as parallel light and the sunlight that is obscured by the Earth produces a cylindrical shadow [19], as shown in Figure 3.

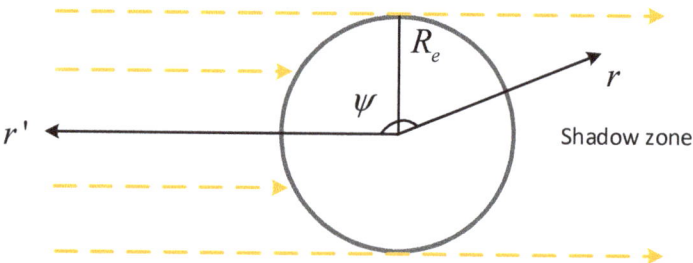

Figure 3. Cylindrical shadow model.

The following formula can be used to discover whether the grid point is in the shadow of the Earth. Satellites are not visible to space objects when they are in the Earth's shadow zone, and satellites are visible to space objects when they are not in the Earth shadow zone. The value of the cylindrical shadow factor is −1 when the space object is in the shadow zone, otherwise it is 1.

$$\begin{cases} -1, \psi > \frac{\pi}{2} \& r < \frac{R_e}{\sin \psi} \\ 1, \text{else} \end{cases} \quad (2)$$

where r is the connection between the grid points and the center of the earth, r' is the connection between the Sun and the center of the Earth, and ψ is the angle between r and r'. The angle ψ can be expressed using the following formula.

$$\cos \psi = \sin \delta_{sun} \sin \delta_{obj} + \cos \delta_{sun} \cos \delta_{obj} \cos(\alpha_{obj} - \alpha_{sun}) \quad (3)$$

where α_{sun} and δ_{sun} are the right ascension and declination of the Sun, α_{obj} and δ_{obj} are the right ascension and declination of the space objects. The position of the Sun changes with the seasons, so it is necessary to obtain multiple coordinate positions of the Sun, calculate the constellation's space coverage performance for each position and identify the average. You can get the average coverage performance of the constellation in a year. In this model, the four positions of the Sun at the spring equinox (0°, 0°), the summer solstice (90°, 23°26′), the autumnal equinox (180°, 0°), and the winter solstice (270°, −23°26′) were taken to calculate the observation performance of the space objects by the constellations at different positions of the Sun.

(4) Visibility function

Define the visual function of F: When the sensor is visible to the grid points, F > 0 is the opposite of F < 0. The sub-apparent function is defined by the two conditions of Formula (1), respectively.

Define the visibility function of F: When the sensor is visible to the grid points, F > 0, and vice versa F < 0. According to the two conditions in Formula (1), the sub-visibility functions are defined separately as shown in the following formula [20].

$$f_1 = \varphi - \beta \tag{4}$$

$$f_2 = \frac{L_{max} - d_{st}}{L_{max}} \tag{5}$$

Define another sub-visibility function based on the Earth's shadow condition as shown in the following formula.

$$f_3 = \begin{cases} -1, \Psi > \frac{\pi}{2} \& r < \frac{R_e}{\sin \Psi} \\ 1, \text{else} \end{cases} \tag{6}$$

It can be seen that when the values of the three Formulas (4)–(6) are greater than zero at the same time, the satellite can observe the grid point.

Define the visibility function as the following.

$$F = \min\{\text{sign}(f_1), \text{sign}(f_2), \text{sign}(f_3)\} \tag{7}$$

Define the visibility for each grid area as its visual value, denoted by C_j, which is 1 if the satellites are visible to the grid, otherwise it is 0. The visual value of each grid can be calculated using Formula (8).

$$C_j = \begin{cases} 1, (F > 0) \\ 0, (F < 0) \end{cases} \tag{8}$$

Visual value calculation rules of the grid: When the same grid can be detected by one satellite in the simulation time, the visual value of the satellite to the grid point is 1. When it is detected by the second satellite, the visual value is 1/2, and so on. When it is detected by the n-th satellite, the visual value is $1/n$.

(5) Constellation period calculation

For a specific grid area, the time interval between the recurrence of the corresponding satellite constellation geometry is the coverage period of the constellation to the grid area, and the smallest repeat interval is called the minimum coverage recurrence period.

The constellation period is calculated as follows.

$$T_C = \frac{T_S P}{T} \tag{9}$$

where $T_s = 2\pi \sqrt{\frac{a^3}{\mu}}$, T_C is the constellation period, T_s is the satellite period, and μ is the Earth's gravity parameter. Here, the value is 398,600.4.

2.2. Theoretical Analysis

The constellation design was based on the walker-δ constellation [21,22]. All the satellite orbits in the constellation were circular orbits, with the same orbital altitude and inclination. The right ascension of all the orbital planes were evenly distributed. Through the establishment of the constellation airspace coverage model, using the appropriate objective functions and constraint conditions and optimizing the design using the genetic algorithm, the best constellation parameters that met the requirements were obtained, and the constellation configuration was determined. This constellation design mainly considered the coverage of the airspace in the orbital altitude range of 300–1000 km.

The assessment of the constellation's coverage performance in the target airspace was measured in terms of time and space. At the same time, the distribution density of space objects in the grid area corresponding to different right ascensions, declinations, and altitudes in the airspace was considered, as shown in Figure 4. The resulting comprehensive coverage of the constellation was the ratio of the statistical sum of the volume, time, and density of all the observable grids of the constellation to the corresponding total.

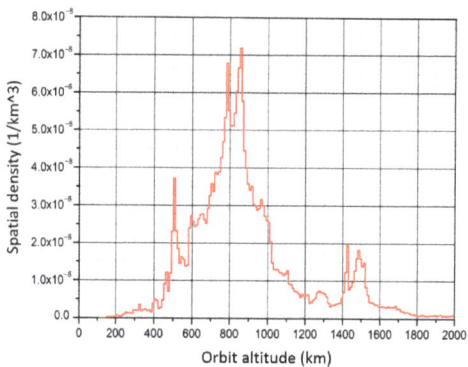

Figure 4. Distribution density of the space objects at different orbital altitudes.

The comprehensive coverage performance of the constellation can be expressed by the following formula, which is also used as the objective function of the constellation's optimization design.

$$M_S = -\frac{\sum_{j=1}^{S}(\rho_j V_j \sum_{l=1}^{T_C} d_t C_j)}{\sum_{j=1}^{S} \rho_j V_j T_C} \quad (10)$$

In the above formula, ρ_j is the spatial debris density corresponding to each grid area, V_j is the volume of each grid area, d_t is the simulation time step, C_j is the apparent value of each grid, S is the total number of grids, T is the number of satellites in the constellation, and T_C is the constellation period.

The constraints are as follows.

$$\begin{aligned} & 12 < T < 24,\ 3 \le P \le 6,\ F = 1 \\ & 0 \le i \le \pi,\ 300\ \text{km} \le a - R_e \le 1000\ \text{km} \\ & 0 \le \Omega_0, \Omega_m \le 2\pi,\ 0 \le M_{0,0}, M_{m,n} \le 2\pi \end{aligned} \quad (11)$$

Finally, the constellation configuration and its parameters were determined as the following parameters: $T/P/F$, a, i, Ω_1, $M_{1,1}$.

Here, T is the number of satellites in the constellation, P is the orbital plane number of the constellation, F is the phase factor, a is the orbital radius of the satellite, i is the orbital inclination, Ω_1 is the right ascension of the ascending nodes of the fiducial satellite in the constellation, $M_{1,1}$ is the mean anomaly of the fiducial satellite, and R_e is the earth radius.

According to the walker-δ constellation configuration, the positions of the satellites in the constellation can be represented by the following formula [23].

$$\begin{cases} \Omega_m = \Omega_1 + (m-1)\frac{360}{P} \\ M_{m,n} = M_{1,1} + (m-1)F\frac{360}{T} + (n-1)P\frac{360}{T} \end{cases} \quad (12)$$

Here, Ω_m is the right ascension of the ascending nodes of the m-th orbital plane in the constellation and $1 \le M \le P$, $M_{m,n}$ is the mean anomaly of the n-th satellite on the m-th orbital plane, $1 \le n \le \frac{T}{P}$.

2.3. Design Results

We established the above constellation design model and optimized it using the genetic algorithms [24,25]. Then the optimal orbit parameters such as orbit altitude and inclination were obtained. According to these orbit parameters, the different constellation configurations were designed, and the spatial and time domain coverage performances of the constellations were analyzed according to the established model to obtain high-quality constellation configurations using screening.

The optimization process was carried out using the genetic algorithm, and the optimization process is shown in Figure 5. It can be seen from the figure that when the population evolved to 55 generations, the optimal solution can be obtained. The optimal orbital altitude was 609.782 km and the inclination angle was 96.142°.

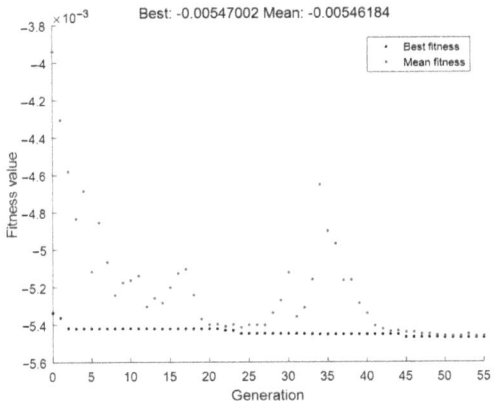

Figure 5. The optimization process of the orbital parameter design using the genetic algorithm.

The following situations were mainly considered, namely the number of orbital planes (P = 3, 4, 5, 6) And the number of constellation satellites (T = 12, 15, 16, 18, 20, 21, 24). Corresponding to the following 13 constellation configurations (12_3P, 12_4P, 12_6P, 15_3P, 15_5P, 16_4P, 18_3P, 18_6P, 20_5P, 21_3P, 24_3P, 24_4P, 24_6P), the specific meaning was that, taking 18_3P as an example, the constellation was composed of 18 satellites that were evenly distributed on three orbital planes, with six satellites on each orbital plane. The right ascension of each orbital plane was evenly distributed and the angle between the adjacent orbital planes was 120°.

According to the optimal deployment orbit altitude and the inclination of the satellite, and then according to the constellation spatial coverage model established above, the spatial coverage performance of each constellation can be calculated to judge the quality of the constellation's monitoring performance.

The calculated airspace coverage performance of each constellation is shown in Figure 6. The airspace coverage performance represents the airspace performance indicators that can be covered by all the different constellations in a constellation cycle, which can be used to compare the performance of each constellation. It can be seen from the figure that with the increase in the number of satellites in the constellation, the airspace coverage performance of the constellation showed an upward trend as a whole. However, the constellation with the same number of satellites had a very different performance due to the different constellation configurations. Therefore, the reasonable design of the constellation configuration plays a very important role for improving the observation efficiency of the constellation.

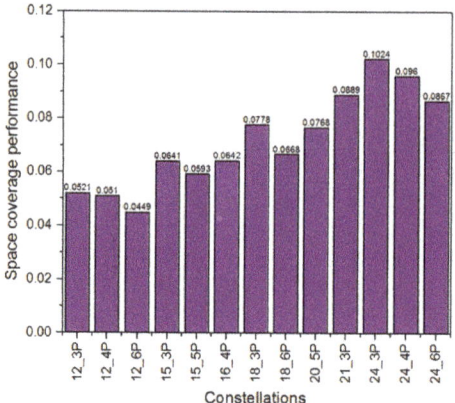

Figure 6. The airspace coverage performance of each constellation.

3. Constellation Observation Optimal Scheduling

3.1. Overview of Task Scheduling

Due to the large number of space objects [26], as shown in Figure 7, the space-borne optical telescope can only track one space object at a time. In order for the satellite constellation to monitor space debris during all weather events and at all times and achieve multi-satellite coordination, make full use of space-based observation equipment resources, reduce inefficient repeated observation, and improve the observation efficiency, it is necessary to develop an optimal scheduling algorithm for the observation tasks of space-based satellite constellations. This includes formulating efficient observation plans for each satellite and improve the monitoring efficiency.

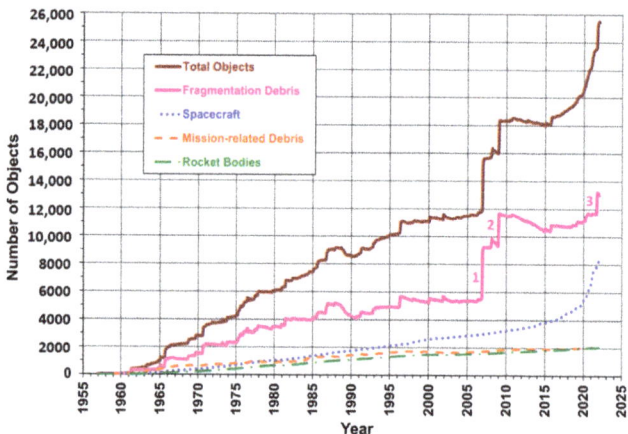

Figure 7. Growth trend in the number of space objects.

The goal of space-based observation equipment detection task scheduling is to allocate detection resources to significantly improve the detection and prediction capabilities for space objects to a certain extent. This allocation must consider many influencing factors, such as the observation capability of each sensor, the orbit accuracy of each observation object, and the detection frequency required to maintain a catalog of space objects. Space-based optical satellite observation task scheduling is a combined optimization problem under complex constraints. Its observation equipment and observation targets move at high speeds; it exhibits multi-time window constraints, multi-resource constraints, and high conflicts; and it has become one of the leading issues in the field of space observation. When the amount of observation equipment and targets increases, the complexity of the optimization problem will increase geometrically.

3.2. Principles of Observation Task Scheduling

Each detection device has its own independent requirements and tasks. Task scheduling is to allocate time for each device to observe the space objects every day. In brief, the satellite network observation task scheduling problem refers to scheduling different satellite observation equipment to observe space objects in the observable time period under certain constraints [27,28]. That is, the detection scheduling result is better under certain expected effects (such as the largest number of observation targets and the highest equipment utilization).

The principle of task optimization scheduling is to track as many space objects as possible [29]. At the same time, the turning angle of the telescope should be made smaller and the idle time of the telescope should be minimized. Observation is given to objects with high priority and the space objects with short arcs are observed first. The objects with small phase angles are also observed first. followed by the targets with small changes, etc.

Suppose there are a total of N space objects to be observed. For any space object $n \in \{1,\ldots,N\}$, the space object is divided into R classes, and the level of each space object is expressed as l_n, $l_n \in \{1,\ldots,R\}$. Each space object can only be observed within a time window, expressed as $[O_n, D_n]$, where O_n is the initial time when the space object can be observed and D_n is the cut-off time.

Suppose there are a total of M optical observation satellites, for which any satellite $m \in \{1,\ldots,M\}$. An optical satellite can observe only one space object at a time, and likewise, each space object can only be assigned to one optical satellite for observation. The definition of the (0, 1) variables $x_{n,m}$, $x_{n,m} = 1$ indicates that the space object to be observed n is assigned to the optical satellite m, otherwise $x_{n,m} = 0$. The observation start time of the space object n is S_n and the observation end time is C_n, which is expressed as the following formula.

$$C_n = \begin{cases} S_n + t_{min}, & \text{if } D_n - S_n \geq t_{min} \\ D_n, & \text{Otherwise} \end{cases} \quad (13)$$

From S_n and C_n, the observation time P_n can be calculated as the following.

$$P_n = C_n - S_n \quad (14)$$

The time it takes for an optical satellite to observe a space object must be $\geq t_{min}$ seconds to be counted as a successful observation. The observation result k_n of the space object n is expressed as the following.

$$k_n = \begin{cases} 1, & P_n \geq t_{min} \\ 0, & \text{Otherwise} \end{cases} \quad (15)$$

where k_n is the (0, 1) variable, $k_n = 1$ indicates that the observation of the space object n is successful, and $k_n = 0$ indicates that the observation failed.

In order to observe as many high-priority space objects as possible, the purpose of observation scheduling is the total weighted number K_{NL} of successful observations of space objects, which can be expressed as the following.

$$K_{NL} = \sum_{n=1}^{N} \sum_{m=1}^{M} x_{n,m} \cdot k_n \cdot l_n \quad (16)$$

Therefore, the constellation observation scheduling model is established as follows.

$$\max K_{NL} = \sum_{n=1}^{N} \sum_{m=1}^{M} x_{n,m} \cdot k_n \cdot l_n \quad (17)$$

The greedy algorithm based on the multi-dimensional list is used for constellation observation task scheduling, and the algorithm is introduced below.

(a) Algorithm Fundamentals

The multi-dimensional dynamic list programming algorithm is a fast heuristic approach for solving the match between the tasks and resources. It is a combination of a set of algorithms, which can be roughly divided into two parts, namely [30,31]:

(1) To determine the priority function of the task.
(2) To determine the priority function of the observation platform.

Among them, the priority function of the observation task is used to clarify that the task needs to be processed and is selected according to the task priority. The priority is selected first, and the subtask execution order is also arranged according to the size of the priority, from largest to smallest, with the large task executed first and the small task executed later. Determining the priority of the satellite platform selection dynamically identifies the priority of each current idle platform according to the platform resource priority function. From all the priority values obtained by all idle platforms, the highest priority is selected first and the subtask is executed, followed by the next highest priority value until the required resources are met for the task to be executed. At the same time, the status of all the satellite platforms assigned to the task is set to the working state until the end of the task processing, and then they are set to an idle state and can be reassigned to a new task.

The algorithm flow chart is shown in Figure 8.

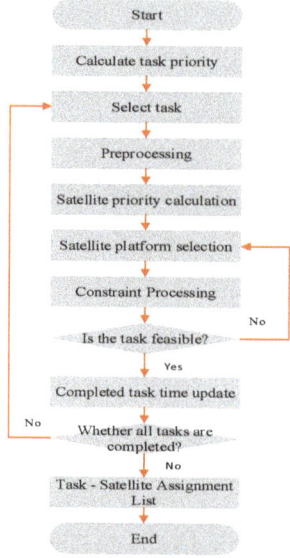

Figure 8. Constellation observation scheduling algorithm flow chart.

3.3. Observation Task Generation

Considering the influence of the various factors on the observation performance of space-based monitoring equipment, we designed an efficient optimal scheduling algorithm. It included the scheduling algorithm based on the priority of space objects and the scheduling algorithm based on the priority of the object update rate. It can realize the optimal scheduling of the observation tasks for a single satellite or a satellite constellation. A large number of space object orbit data are used for the numerical simulation analysis, and the analysis results verify the feasibility and adaptability of the scheduling method.

The observation task optimization algorithm we designed can create an observation plan for a single satellite or a constellation composed of multiple satellites at the same time. The observation objects can be set to prioritize the observation of key objects. The performance parameters of the satellite detectors can be set, and the number of observation satellites can be increased or reduced as needed.

The overall consideration for the optimization scheduling algorithm of the observation tasks is to observe as many objects as possible, and simultaneously make the rotation angle of each satellite switch smaller and observe the higher priority objects first. The priority of a single arc is higher than that of multiple arcs, considering the conditions of the sky, light, and Earth shadows, such as the Moon and the Sun.

4. Cataloging Requirements

The cataloging of space objects means that the monitoring network can continuously observe, match, correlate, and determine the orbits of space objects under normal operating conditions and can update the orbit information in a timely manner for each space object in the catalog. The number of space objects in the catalogue is an important index that reflects the ability of the monitoring network. The system performance is directly related to the number of space objects that are correctly cataloged.

The conditions for the space objects being judged to be correct for cataloging were as follows. The maximum observable time interval of LEO objects was about 24–48 h, the maximum observable time interval of GEO objects was about 48–168 h, and the maximum observable time interval o other space objects was less than 168 h. If the vast majority of LEO space objects can be observed every 24 h, the orbital cataloguing for most LEO space objects can be achieved. In Section 5, the monitoring of space objects below 1000 km orbital altitude within 24 h on 26 February 2022 was calculated by the simulation, which was used to measure the cataloging ability of LEO space objects that were to be measured by the constellation.

The calculation process was as follows: (1) Calculate the observation time interval for each space object. (2) Calculate the maximum revisit time interval for each space object. (3) Calculate the maximum observation time interval distribution function for the entire set of observation objects. (4) Calculate the percentage of the time interval for cataloging space objects that are less than the value defined above.

5. Numerical Simulation Experiment

5.1. Experiment Overview

The simulation model of the detection equipment was established by comprehensively considering the aperture of the observation telescope and limiting the detection magnitude; space object orbit data; sky, light, and Earth shadow conditions; the space environment model; and atmospheric influence, etc. The number of space objects, arc segments, and detection frequencies that can be detected were numerically simulated and analyzed. The observation ability of the space objects with a different number of observation satellites and different constellation configurations was simulated and calculated. The number of objects and arcs that could be detected under each constellation configuration, as well as the distribution of the arc length and detection frequency, were analyzed, and the detection performance for several constellation configurations was compared.

The time of the simulation analysis was 26 February 2022, and the simulation duration was 24 h. The simulation was mainly for space objects with an orbital altitude less than 1000 km. It was assumed that the orientation of the detector on the satellite could be rotated according to the mission requirements during observation to realize the tracking and observation of space objects.

The simulation was mainly divided into two parts. The first part simulated and calculated the situation of the space objects passing through the detection range of the satellite detectors under different constellation configurations. That is, it analyzed the basic information, such as the number of objects and arcs, the arc length distribution, detectable frequency, and so on in order to understand the influence of the orbital position of each constellation on the observation performance of the space objects. The second part was mainly based on the observation task scheduling algorithm of the space-based optical satellite constellation established above to allocate and schedule the observation tasks for the different constellation configurations. The various satellites in the constellation could coordinate, cooperate, and make full use of the observation resources and observation time. More space objects can be observed, and more observation data can be obtained, thereby providing support for the determination of the orbit of space objects and the establishment of a library of orbit catalogues of space objects.

For convenience, each constellation was numbered, as shown in Table 1.

Table 1. Constellation number table.

Constellation	Meaning
Constellation 1	12_3P, 12 satellites were evenly distributed on three orbital planes
Constellation 2	12_4P, 12 satellites were evenly distributed on four orbital planes
Constellation 3	12_6P, 12 satellites were evenly distributed on six orbital planes
Constellation 4	15_3P, 15 satellites were evenly distributed on three orbital planes
Constellation 5	15_5P, 15 satellites were evenly distributed on five orbital planes
Constellation 6	16_4P, 16 satellites were evenly distributed on four orbital planes
Constellation 7	18_3P, 18 satellites were evenly distributed on three orbital planes
Constellation 8	18_6P, 18 satellites were evenly distributed on six orbital planes
Constellation 9	20_5P, 20 satellites were evenly distributed on five orbital planes
Constellation 10	21_3P, 21 satellites were evenly distributed on three orbital planes
Constellation 11	24_3P, 24 satellites were evenly distributed on three orbital planes
Constellation 12	24_4P, 24 satellites were evenly distributed on four orbital planes
Constellation 13	24_6P, 24 satellites were evenly distributed on six orbital planes

5.2. Object Detectable Analysis

5.2.1. Detection Quantity Analysis

As shown in Figure 9, the observation performance of space objects varies greatly among the different constellation configurations, and the observation efficiency of the constellations 12_3, 12_4, and 12_6 that were composed of 12 satellites was relatively poor. Among them, the 12_3 constellation configuration meant that 12 satellites were evenly distributed on three orbital planes, and the right ascension of each orbital plane was 120 degrees different. The meaning of other constellation configurations can be deduced in the same manner.

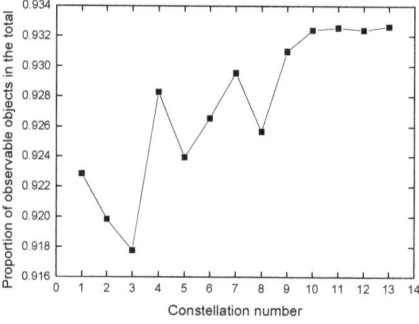

Figure 9. The proportion of space objects within the detectable range of the different constellation configurations to the total.

Secondly, the constellations composed of 15, 16, and 18 satellites had good monitoring performances, of which the observation performances of the 15_3 and 18_3 satellites were relatively good. In addition, the constellations composed of 20, 21, and 24 satellites had the best performances and the largest numbers of observed targets. However, the larger the constellation size, the greater the cost. Therefore, the specific constellation to be selected for the space object observation should be comprehensively considered according to the demand and the cost.

5.2.2. Analysis of the Observation Arc Length

Figure 10 shows the relationship between the number of observable space objects in the different constellations and the length of the arcs. It can be seen that there were various arc lengths in the space objects observed by the different constellation configurations. For the same range of the detection arc length, the overall trend was that the more satellites the constellation has, the more space objects it can observe. However, for constellations with the same number of satellites, the monitoring efficiency was also slightly different due to the different constellation configurations. According to the post-processing requirements of the monitoring data, if the arc length was long, it was conducive to the determination of the space object orbit and the improvement in the accuracy.

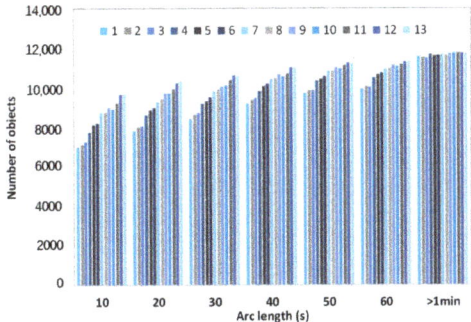

Figure 10. The distribution of observable objects with an arc length in the different constellation configurations.

Figure 11 shows the distribution of the number of arcs and arcs lengths observed in the different constellation configurations. It can be seen that the number of arcs corresponding to the different arc lengths was quite different. Among them, the number of arcs with arc lengths greater than 1 min was large, which was also conducive to improving the effectiveness of the observation data. Long observation arc lengths are preferred for space objects, which is conducive to data processing and the orbit determination of space objects. When the arc length was less than 1 min, it can be seen from the above figure that the number of arcs corresponding to the arc length gradually increased from 10 s to 60 s. For the same arc length range, for example, when the arc length was approx. 40 s, the larger the constellation size, the more arc segments could be observed.

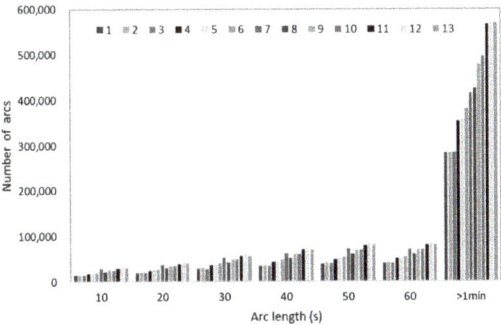

Figure 11. Observable arc length distribution of the different constellation configurations.

5.2.3. Observation Frequency

Figure 12 shows the relationship between the number of observable space objects in the different constellations and the detectable frequency of these objects. As can be seen from the figure, for the constellations with a small number of satellites, the detectable frequency for most space objects was approx. 30–50 times. For large-scale constellations with more satellites, the detectable frequency of space objects increased considerably by approx. 70–80 times. Therefore, the larger the constellation size, the more observation opportunities, and the easier it is to observe the key targets.

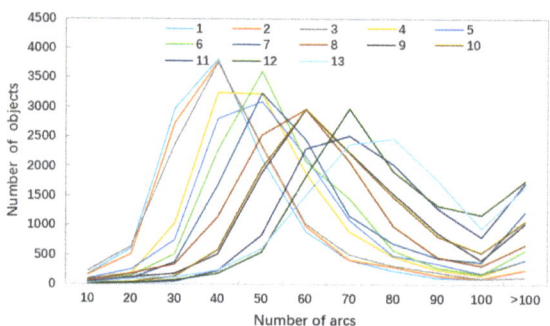

Figure 12. Variation of the number of objects observed using the different constellation configurations for the observation frequency.

5.3. Tacking Observation Performance

Based on the optimal scheduling algorithm of the satellite constellation observation tasks designed above, the monitoring ability of the different constellation configurations for LEO space objects was simulated, mainly for space objects with orbital altitudes less than 1000 km. The simulation conditions were that each object was observed for at least 60 s, and the switching time of the satellite detectors was 5 s. The field of view of the telescope was 15°, the maximum observation distance was 1000 km, and the simulation duration was 24 h.

It can be seen from Figure 13 that the fourth, seventh, and eleventh constellation configurations had good tracking observation performances. The corresponding constellation configurations were 15_3, 18_3, and 24_3, respectively. They tracked 91.10%, 91.77%, and 92.75% of LEO space objects, respectively, within 24 h. If they were observed continuously every day, most LEO space objects could be catalogued. Among these constellation configurations, 15_3 and 18_3 were the locally optimal constellation configurations, and the 24_3 constellation was the globally optimal constellation configuration. However, for the constellations composed of 15 and 18 satellites, the number of satellites was less than the 24-satellite constellation, resulting in a lower cost and a higher cost performance. The specific constellation configuration can be determined according to the needs of the users.

In Figure 14, the 15_3 constellation configuration was taken as an example, to analyze the number of space objects that could be observed by each satellite in the constellation under the observation task optimization scheduling algorithm. Here, 15_3 indicated that the constellation was composed of 15 satellites, which were evenly distributed on three orbital planes, and the right ascension of each orbital plane was evenly distributed. The detector on the satellite could rotate freely, so that it could observe the specified space objects. The number of space objects that could be observed by each satellite in the constellation within 24 h is shown in the above figure. It can be seen that the number of space objects that can be tracked by most satellites was more than 800, and the number of objects that can be tracked by a few satellites was approx. 650. The constellation tracked and observed 11,522 objects in 24 h, accounting for 91.10% of the total objects participating in the simulation. It had a good observation efficiency. If the constellation continued to observe space debris every day, it could catalog most space objects with an orbital height of less than 1000 km.

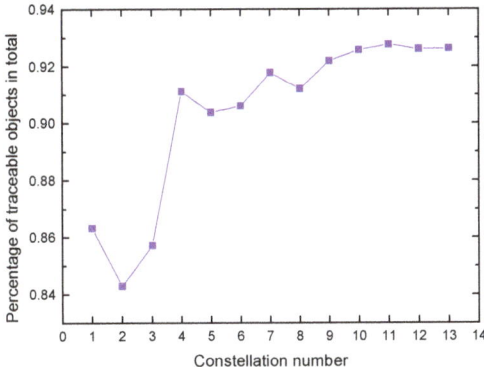

Figure 13. Proportion of objects that can be tracked within 24 h in the different constellation configurations.

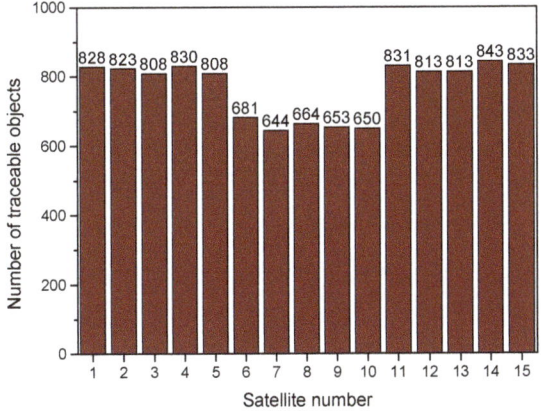

Figure 14. The number of objects that could be tracked by each satellite in the 15_3 constellation in 24 h.

6. Conclusions

In this paper, a mathematical model of a constellation design was established by considering various factors, and was optimized by using a genetic algorithm to obtain the optimal orbital height of 609.782 km, the optimal orbital inclination angle of 96.142°, and a high-quality constellation configuration, which was mainly used to observe space objects with an orbital altitude below 1000 km. A greedy algorithm based on a multi-dimensional list was designed to optimize the scheduling of the constellation observation tasks and realize the optimization of the single-satellite observation tasks and constellation observation tasks. Through the simulation experiments, the mathematical model and the optimization algorithm established above were verified.

It was found that the constellation composed of 15 satellites had a high cost performance, achieved a high observation efficiency with relatively few satellites, and tracked 91% of the objects with an orbital altitude of less than 1000 km every 24 h. The constellation composed of 24 satellites distributed on three orbital planes had the best observation efficiency. It tracked approx. 93% of LEO orbit objects in 24 h. Through long-term continuous observation, these constellations could maintain an orbit catalogue of most low orbit objects, identify dangerous rendezvous times, and prevent the loss of space assets.

Author Contributions: Conceptualization, G.L.; methodology, G.L., J.L. and H.J.; software, G.L.; validation, G.L., J.L. and H.J.; writing—original draft preparation, G.L.; writing—review and editing, G.L., J.L., H.J. and C.L.; supervision, J.L., H.J. and C.L.; project administration, J.L.; funding acquisition, J.L. and H.J. All authors have read and agreed to the published version of the manuscript.

Funding: This research was funded by the Space Debris Research Project of China (No. KJSP2020020201, No. KJSP2020020202), the China National Key R&D Program during the 14th Five-year Plan Period (No. 2022YFC2807304), and the Foundation Laboratory of Pinghu (No. 2020AB01001).

Institutional Review Board Statement: Not applicable.

Informed Consent Statement: Not applicable.

Data Availability Statement: Not applicable.

Conflicts of Interest: The authors declare no conflict of interest.

References

1. Klinkrad, H. *Space Debris: Models and Risk Analysis*; Springer: New York, NY, USA, 2006.
2. Olmedo, E.; Sanchez-Ortiz, N. Space debris cataloguing capabilities of some proposed architectures for the future european space situational awareness system. *Mon. Not. R. Astron. Soc.* **2010**, *403*, 253–268. [CrossRef]
3. Riot, V.; de Vries, W.; Simms, L.; Bauman, B.; Carter, D.; Phillion, D.; Olivier, S. The Space_based Telescopes for Actionable Refinement of Ephemeris (STARE) mission. In Proceedings of the 27th Annual AIAA/USU Conference on Small Satellites, Logan, UT, USA, 10–15 August 2013.
4. Li, G.; Liu, J.; Cheng, H. Space Debris Laser Ranging Technology and Applications. *Space Debris Res.* **2020**, *20*, 40–48.
5. Cui, Z.; Xu, Y. Impact simulation of Starlink satellites on astronomical observation using worldwide telescope. *Astron. Comput.* **2022**, *41*, 100652. [CrossRef]
6. Julian, R.; Emiliano, C.; Thomas, S. The stare and chase observation strategy at the Swiss Optical Ground Station and Geodynamics Observatory Zimmerwald: From concept to implementation. *Acta Astronaut.* **2021**, *189*, 352–367.
7. Lawrence, A.; Rawls, M.L.; Jah, M.; Boley, A.; Di Vruno, F.; Garrington, S.; McCaughrean, M. The case for space environmentalism. *Nat. Astron.* **2022**, *6*, 428–435. [CrossRef]
8. Ravago, N.; Jones, B.A. Risk-aware sensor scheduling and tracking of large constellations. *Adv. Space Res.* **2021**, *68*, 2530–2550. [CrossRef]
9. Steindorfer, M.A.; Kirchner, G.; Koidl, F.; Wang, P.; Jilete, B.; Flohrer, T. Daylight space debris laser ranging. *Nat. Commun.* **2020**, *11*, 3735. [CrossRef]
10. Pierre, B.; Marc, D.; Georges, Z. Large satellite constellations and space debris: Exploratory analysis of strategic management of the space commons. *Eur. J. Oper. Res.* **2022**, *304*, 1140–1157. [CrossRef]
11. Li, G.; Jiang, H.; Cheng, H.; Liu, J. Research on monitoring effectiveness of optical satellite constellation. In Proceedings of the 68th International Astronautical Congress (IAC), Adelaide, Australia, 25–29 September 2017.
12. Ma, B.; Shang, Z.; Hu, Y.; Hu, K.; Wang, Y.; Yang, X.; Jiang, P. Night-time measurements of astronomical seeing at Dome A in Antarctica. *Nature* **2020**, *583*, 771–774. [CrossRef]
13. Huang, Y.; Mu, Z.; Wu, S.; Cui, B.; Duan, Y. Revising the Observation Satellite Scheduling Problem Based on Deep Reinforcement Learning. *Remote Sens.* **2021**, *13*, 2377. [CrossRef]
14. Silha, J.; Schildknecht, T.; Hinze, A.; Flohrer, T.; Vananti, A. An optical survey for space debris on highly eccentric and inclined meo orbits. *Adv. Space Res.* **2017**, *59*, 181–192. [CrossRef]
15. López-Casado, C.; Pérez-del-Pulgar, C.; Muñoz, V.F.; Castro-Tirado, A.J. Observation scheduling and simulation in a global telescope network. *Future Gener. Comput. Syst.* **2019**, *95*, 116–125. [CrossRef]
16. Ip, A.W.; Xhafa, F.; Dong, J.; Gao, M. Chapter 6—An overview of optimization and resolution methods in satellite scheduling and spacecraft operation: Description, modeling, and application. In *Aerospace Engineering, IoT and Spacecraft Informatics*; Elsevier: Amsterdam, The Netherlands, 2022; pp. 157–217; ISBN 9780128210512.
17. Jing, L.; Hong, C. Deployment and Performance Analysis of Early-Warning Satellite Based on STK. *Shipboard Electron. Countermeas.* **2012**, *35*, 1–4.
18. Wang, C.; Chen, X.; Deng, Y. Infrared LEO Constellation Design by GDE 3 Algorithm. *J. Beijing Univ. Aeronaut. Astronaut.* **2010**, *36*, 857–862.
19. Zhang, R.; Tu, R.; Zhang, P.; Liu, J.; Lu, X. Study of satellite shadow function model considering the overlapping parts of Earth shadow and Moon shadow and its application to GPS satellite orbit determination. *Adv. Space Res.* **2019**, *63*, 2912–2929. [CrossRef]
20. Zhang, S.; Zhu, Z.; Hu, H.; Li, Y. Research on Task Satellite Selection Method for Space Object Detection LEO Constellation Based on Observation Window Projection Analysis. *Aerospace* **2021**, *8*, 156. [CrossRef]
21. Guan, M.; Xu, T.; Gao, F.; Nie, W.; Yang, H. Optimal Walker Constellation Design of LEO-Based Global Navigation and Augmentation System. *Remote Sens.* **2020**, *12*, 1845. [CrossRef]

22. Jia, L.; Zhang, Y.; Yu, J.; Wang, X. Design of Mega-Constellations for Global Uniform Coverage with Inter-Satellite Links. *Aerospace* **2022**, *9*, 234. [CrossRef]
23. Zhang, Y.; Fan, L.; Zhang, Y.; Xiang, J.H. *Theory and Design of Satellite Constellations*; Science Press: Beijing, China, 2008.
24. Hu, M.; Ruan, Y.; Zhou, H.; Xu, J.; Xue, W. Long-Term Orbit Prediction and Deorbit Disposal Investigation of MEO Navigation Satellites. *Aerospace* **2022**, *9*, 266. [CrossRef]
25. Yasini, T.; Roshanian, J.; Taghavipour, A. Improving the low orbit satellite tracking ability using nonlinear model predictive controller and Genetic Algorithm. *Adv. Space Res.* **2023**, *71*, 2723–2732. [CrossRef]
26. Orbital Debris Quarterly News. National Aeronautics and Space Administration. Volume 26, Issue 1; March 2022. Available online: https://orbitaldebris.jsc.nasa.gov (accessed on 12 August 2022).
27. Chen, X.; Reinelt, G.; Dai, G.; Wang, M. Priority-based and conflict-avoidance heuristics for multi-satellite scheduling. *Appl. Soft Comput.* **2018**, *69*, 177–191. [CrossRef]
28. Ben-Larbi, M.K.; Pozo, K.F.; Haylok, T.; Choi, M.; Grzesik, B.; Haas, A.; Krupke, D.; Konstanski, H.; Schaus, V.; Fekete, S.P.; et al. Towards the automated operations of large distributed satellite systems. Part 1: Review and paradigm shifts. *Adv. Space Res.* **2021**, *67*, 3598–3619. [CrossRef]
29. Liu, Y.; Zhang, S.; Hu, H. A conflict avoidance algorithm for space-based collaborative stereo observation mission scheduling of space debris. *Adv. Space Res.* **2022**, *70*, 2302–2314. [CrossRef]
30. Chang, Z.; Chen, Y.; Yang, W.; Zhou, Z. Mission planning problem for optical video satellite imaging with variable image duration: A greedy algorithm based on heuristic knowledge. *Adv. Space Res.* **2020**, *66*, 2597–2609. [CrossRef]
31. Qu, G.; Brown, D.; Li, N. Distributed Greedy Algorithm for Satellite Assignment Problem with Submodular Utility Function**The work is supported by Lincoln Laboratory with award #7000292526. *IFAC-PapersOnLine* **2015**, *48*, 258–263. [CrossRef]

Disclaimer/Publisher's Note: The statements, opinions and data contained in all publications are solely those of the individual author(s) and contributor(s) and not of MDPI and/or the editor(s). MDPI and/or the editor(s) disclaim responsibility for any injury to people or property resulting from any ideas, methods, instructions or products referred to in the content.

Article

Non-Cooperative Spacecraft Pose Measurement with Binocular Camera and TOF Camera Collaboration

Liang Hu [1,2,3], Dianqi Sun [1,2,3], Huixian Duan [1,3], An Shu [1,3], Shanshan Zhou [1,3] and Haodong Pei [1,3,*]

1. Shanghai Institute of Technical Physics, Chinese Academy of Sciences, Shanghai 200083, China
2. University of Chinese Academy of Sciences, Beijing 100049, China
3. Key Laboratory of Intelligent Infrared Perception, Chinese Academy of Sciences, Shanghai 200083, China
* Correspondence: peihaodong@sina.com

Abstract: Non-cooperative spacecraft pose acquisition is a challenge in on-orbit service (OOS), especially for targets with unknown structures. A method for the pose measurement of non-cooperative spacecrafts based on the collaboration of binocular and time-of-flight (TOF) cameras is proposed in this study. The joint calibration is carried out to obtain the transformation matrix from the left camera coordinate system to the TOF camera system. The initial pose acquisition is mainly divided into feature point association and relative motion estimation. The initial value and key point information generated in stereo vision are yielded to refine iterative closest point (ICP) frame-to-frame registration. The final pose of the non-cooperative spacecraft is determined through eliminating the cumulative error based on the keyframes in the point cloud process. The experimental results demonstrate that the proposed method is able to track the target spacecraft during aerospace missions, which may provide a certain reference value for navigation systems.

Keywords: space traffic management; binocular camera; TOF camera; joint calibration; non-cooperative spacecraft; pose estimation

1. Introduction

The data show that nearly 2000 objects larger than 10 cm in diameter have been found in low-Earth orbit [1]. Debris not only occupies valuable orbital resources but also increases the risk of collision with orbiting satellites, resulting in the removal of space debris becoming an urgent problem [2–4]. The concept of OOS was proposed in the 1960s, and a great deal of research work has been carried out in this field, with more than 130 missions launched [5–8]. Since targets are in a state of free tumbling in space [9], it is necessary to determine the attitude of the target spacecraft in real time to provide accurate information for the guidance, navigation, and control (GNC) system [10], which is an indispensable condition for the acquisition of non-cooperative targets.

Monocular cameras, binocular cameras, lidar and TOF cameras are commonly used sensors in short-range detection [11]. Acquiring non-cooperative target poses by means of stereo vision has been extensively studied. Numerous studies have accomplished pose estimation by identifying docking rings and other features of satellites [12–15]. The disadvantage of an optical camera is that its imaging is greatly affected by illumination, which has certain limitations. The method based on point cloud estimation can overcome the influence of space's complex environment. Liu et al. proposed a pose estimation method based on the known target model and the point cloud data generated by the lidar. The main advantage of this method is that the point cloud data are processed directly without the detection and tracking of features [16]. Opromolla et al. designed an approach combining the principal component analysis and template matching. The authors focused on its ability to succeed in the measurement task without any initial guess work [17]. Guo et al. presented a pose initialization based on template matching with sparse point cloud input, mainly concentrating on offline template construction [18]. However, the premise of these methods requires the CAD model on the ground experiment to achieve the pose measurement of the target satellite. In addition, due to the need for a good initial value in the point cloud registration and a weaker echo signal when the target moves at a higher speed, it is difficult to solve the problem of pose estimation based on point clouds. It is a challenge for a single sensor to reliably track a target in real time, which demonstrates the necessity of studying the pose measurement method based on multisensor collaboration.

In terms of active and passive means of collaborative navigation, Terui et al. combined stereo vision and the ICP algorithm to estimate the motion of space debris. Their work mainly focused on using time series images to cope with the disadvantages of the ICP algorithm [19]. Peng et al. designed a method to simply fuse the point cloud data reconstructed by stereo vision and the point cloud scanned through the laser radar. The extended Kalman filter algorithm was generated to acquire the pose and velocity of the non-cooperative target. However, the real-time performance needed to be improved [20]. Guo et al. proposed a target recognition algorithm based on information fusion with binocular vision and laser radar, and the attitude measurement's accuracy was improved through a simulation experiment; however, the simulation model in this article is too simple [21]. Liu et al. proposed an accurate pose estimation method for a non-cooperative target based on a TOF camera coupled with a grayscale camera. This major work is based on 2D line and 3D line correspondence. However, some salient feature points of the target model are not considered [22]. Su et al. presented a pose tracking method for on-orbit uncooperative targets based on the deep fusion of an optical camera and laser. The authors concentrated on acquiring a dense point cloud with scale information. However, the author does not mention the elimination of cumulative errors in point cloud registration [23].

The excellence of a grayscale camera is that it can extract the details of the target with less computational effort. Active means, such as point cloud, require a large amount of computation when extracting relevant key features. Based on the advantage of two sensors, a pose estimation method based on stereo vision and point cloud tracking is proposed in this paper. The proposed method can settle the problem of initial pose acquisition and does not depend on a CAD model during spacecraft tracking. A new joint calibration method is proposed to ensure the minimum error in the conversion between different coordinate systems. A new feature point association criterion is also proposed in Section 3.2.1. The information generated in the initial pose acquisition is mapped to refine the frame-to-frame point cloud registration. Two kinds of motion experiments were performed, namely, high speed and low speed, in order to verify the feasibility of the proposed method. This method can be applied to the mission of the chaser satellite navigation systems in OOS.

The rest of this paper is organized as follows. Section 2 defines the problem of non-cooperative spacecraft pose measurement. Section 3 describes in detail the pipeline of the proposed pose estimation method. Section 4 comprises the analysis of the calibration results and the low- and high-speed experiments' results. Section 5 summarizes the full text.

2. Problem Definition

Figure 1 presents a schematic diagram of capturing non-cooperative targets in aerospace missions. Accurate pose data are provided by the navigation system. The capturing task is completed by the robot arm at the appropriate position.

A schematic diagram of the composite pose measurement system built in this paper is shown in Figure 2. The visual image is acquired by grayscale cameras, and the point cloud data are acquired by a TOF camera. The projection points of the space points $P(X_w, Y_w, Z_w)$ in the world coordinate system on the left and right camera imaging planes can be expressed as $P_l(u_l, v_l)$ and $P_r(u_r, v_r)$, respectively. The left and right camera coordinate systems are $O_l - X_l Y_l Z_l$ and $O_r - X_r Y_r Z_r$, respectively. The TOF camera coordinate system is $O_p - X_p Y_p Z_p$.

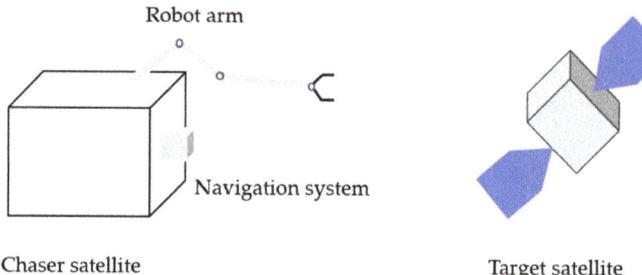

Figure 1. Schematic diagram of the capture mission.

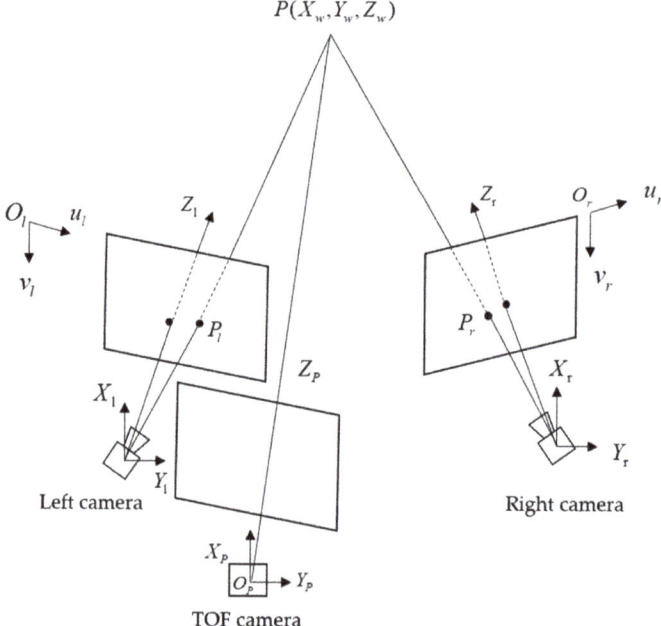

Figure 2. Schematic diagram of the pose measurement system.

The final pose of the non-cooperative target can be determined by Equation (1).

$$\begin{bmatrix} X_p \\ Y_p \\ Z_p \end{bmatrix} = [R|T] \begin{bmatrix} X_w \\ Y_w \\ Z_w \end{bmatrix} \quad (1)$$

where T is the translation vector of the target spacecraft. The rotation matrix, R, of non-cooperative targets in the world coordinate system can be represented by the following formula:

$$R = R_Z(\gamma)R_Y(\beta)R_X(\alpha) = \begin{bmatrix} r_{11} & r_{12} & r_{13} \\ r_{21} & r_{22} & r_{23} \\ r_{31} & r_{32} & r_{33} \end{bmatrix} \quad (2)$$

Then, the Euler angle of the three axes can be calculated by Equation (3).

$$\begin{cases} \alpha = \arctan(r_{32}, r_{33}) \\ \beta = \arctan(-r_{31}, \sqrt{r_{32}^2 + r_{33}^2}) \\ \gamma = \arctan(r_{21}, r_{11}) \end{cases} \quad (3)$$

where α, β, γ represent the X, Y, and Z three-axes angle transformation of the target. $r_{ij}(i = 1, 2, 3; j = 1, 2, 3)$ are the corresponding elements in the rotation matrix. The essence of solving the pose relationship of non-cooperative targets is the calculation of R and T.

3. Detailed Description of the Proposed Method

The specific details of the pose measurement method are presented in this section. Figure 3 illustrates the general flow of the proposed method in this study.

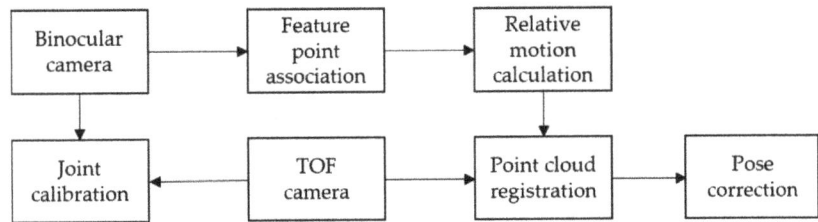

Figure 3. Flow chart of the pose measurement algorithm for non-cooperative spacecrafts.

3.1. System Joint Calibration

Binocular system calibration is first required. Points in the left and right camera coordinate system and the world coordinate system can be represented as X_l, X_r and X_w, respectively. Then, the following formula is:

$$\begin{cases} X_l = R_l X_w + T_l \\ X_r = R_r X_w + T_r \end{cases} \quad (4)$$

Combined with the above formula, there are:

$$X_r = [R_r R_l^{-1}] X_l + [T_r - R_r R_l^{-1} T_l] \quad (5)$$

$$\begin{cases} R_{rl} = R_r R_l^{-1} \\ T_{rl} = T_r - R_r R_l^{-1} T_l \end{cases} \quad (6)$$

where R_l, T_l and R_r, T_r represent the rotation matrix and translation vector from the camera coordinate system to the world coordinate system, respectively. R_r, T_r are the external parameters of the left and right camera that need to be calculated in the binocular calibration.

The feasibility of a circular calibration board was verified in our previous work [24]. The circular calibration board is used to replace the traditional checkerboard calibration board. The center coordinates of the circle can be accurately extracted from the image. The circular calibration board is moved to different positions to acquire images that fill the entire camera field of view. Figure 4 shows the extraction results of the circular marker points on the left and right camera calibration boards.

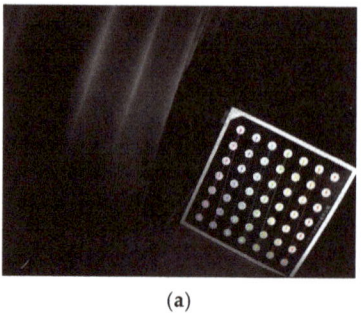
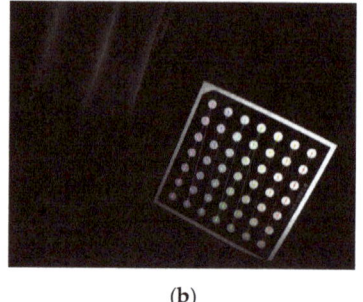

Figure 4. The extraction results of the circle center of the calibration plate in the left and right cameras: (**a**) left image; (**b**) right image.

After completing the calibration of the binocular system, we must perform the joint calibration of the pose measurement system. The purpose of this step is to obtain the transformation relationship between the camera coordinate system and the point cloud coordinate system. It is impossible to acquire accurate three-dimensional coordinates of the circle center directly in the point cloud. Therefore, the corresponding center coordinates are yielded based on the intensity image of the TOF camera, and we map them to the point cloud coordinate system. Figure 5 illustrates the circle extraction result in the intensity image.

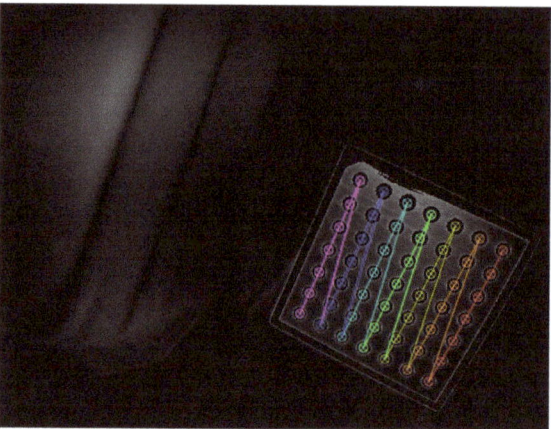

Figure 5. The extraction result of the circle center of the calibration plate in the intensity image.

For the 2D points (u_l, v_l) in the left camera image coordinate system and the corresponding 3D point (X_p, Y_p, Z_p) in the point cloud coordinate system, the following relationship holds true:

$$Z_c \begin{bmatrix} u_l \\ v_l \\ 1 \end{bmatrix} = R_{kl} \begin{bmatrix} R_{pc} & T_{pc} \end{bmatrix} \begin{bmatrix} X_p \\ Y_p \\ Z_p \\ 1 \end{bmatrix} = A \begin{bmatrix} X_p \\ Y_p \\ Z_p \\ 1 \end{bmatrix} \qquad (7)$$

where $R_{kl} = \begin{bmatrix} f_{xl} & 0 & c_{xl} \\ 0 & f_{yl} & c_{yl} \\ 0 & 0 & 1 \end{bmatrix}$, R_{pc} and T_{pc} represent the joint calibration relationship to be solved. The A matrix can be represented as $A = \begin{bmatrix} a_{11} & a_{12} & a_{13} & a_{14} \\ a_{21} & a_{22} & a_{23} & a_{24} \\ a_{31} & a_{32} & a_{33} & a_{34} \end{bmatrix}$. The joint calibration problem becomes the problem of solving matrix A. The initial iteration value of matrix A is yielded based on the direct linear transformation (DLT). According to Formula (7):

$$\begin{bmatrix} -P & u_1 P \\ -P & v_1 P \end{bmatrix} \begin{bmatrix} H_1 \\ H_2 \\ H_3 \end{bmatrix} = 0 \quad (8)$$

where $H_1 = [a_{11}, a_{12}, a_{13}, a_{14}]$, $H_2 = [a_{21}, a_{22}, a_{23}, a_{24}]$, $H_3 = [a_{31}, a_{32}, a_{33}, a_{34}]$, $P = [X_p, Y_p, Z_p, 1]$.

Formula (9) is then acquired:

$$\begin{bmatrix} -X_p & -Y_p & -Z_p & -1 & 0 & 0 & 0 & 0 & u_1 X_p & u_1 Y_p & u_1 Z_p & u_1 \\ 0 & 0 & 0 & 0 & -X_p & -Y_p & -Z_p & -1 & v_1 X_p & v_1 Y_p & v_1 Z_p & v_1 \end{bmatrix} \cdot B^T = 0 \quad (9)$$

where $B = [a_{11} \ a_{12} \ a_{13} \ a_{14} \ a_{21} \ a_{22} \ a_{23} \ a_{24} \ a_{31} \ a_{32} \ a_{33} \ a_{34}]$.

Since there are twelve variables in matrix A, at least six pairs of matching points are required to achieve the linear solution to all variables. The objective function is constructed based on the sum of the reprojection errors of the three-dimensional circle center coordinates, which can be expressed as:

$$E(R_{pc}, T_{pc}) = \operatorname{argmin} \sum_{i=1}^{n} \left\| z_1^i - \frac{1}{s_i} R_{kl}(R_{pc} P_i + T_{pc}) \right\|_2^2 \quad (10)$$

where $z_1^i = [u_1^i, v_1^i]$, s_i is the homogeneous coefficient in the process of coordinate transformation, i.e., the circular center depth, Z_c. $\|\cdot\|_2$ represents the L2 norm of the matrix. The Levenberg–Marquardt (LM) algorithm [25] is used to optimize the reprojection function to minimize the error function and obtain the optimal solution of the joint calibration.

3.2. Binocular Initial Pose Acquisition

The main purpose of this part is to approximately gain the motion state of the target spacecraft and provide a more accurate relative initial value for the subsequent point cloud pose estimation.

3.2.1. Feature Point Association

Common feature point extraction methods include the scale-invariant feature transform (SIFT) operator, speeded-up robust features (SURF) operator, and oriented FAST and rotated BRIEF (ORB) operator [26]. Compared with the ORB operator, the SURF operator has better robustness. Compared with the SIFT operator, it has the advantages of easier real-time processing and implementation. Overall, the SURF operator was implemented to extract the feature points in this paper. The SURF feature point response value is defined in the feature point extraction process, which represents the robustness of the feature point. The point with a small response value will be removed by calculating the response values of the feature points and sorting them.

A relatively complete feature point association criterion is proposed when tracking the target satellite in our study. As presented in Figure 6, a diagram of the feature point extraction and tracking is proposed.

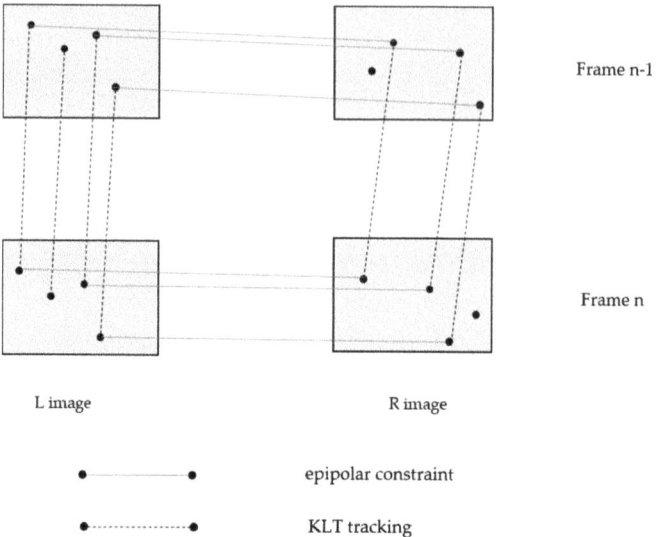

Figure 6. Schematic diagram of the feature extraction and tracking.

The optical flow and epipolar constraint method are combined to track the feature points at the front and back moments in order to ensure the accuracy of the feature point association. In addition, the depth map obtained by the TOF camera is introduced to guide the feature point association. The depth map is projected to the left camera coordinate system through the joint calibration in Section 3.1. The disparity map of the TOF camera can be calculated with Equation (11):

$$D_T = \frac{f_l}{Z_T} \cdot B \quad (11)$$

where D_T is the disparity of the TOF camera, Z_T represents the depth value, f_l is the focal length of the left camera, and B represents the baseline of the camera system.

Finally, the feature points $P_l(u_l, v_l)$ and $P_r(u_r, v_r)$ matched by the left and right cameras must meet the following condition:

$$|D_l(p) - D_T(p)| \leq T_1 \quad (12)$$

where $D_l(p) = u_l - u_r$, which represents the disparity of the left camera. T_1 is usually set to 2. The threshold represents the disparity consistency of the feature points.

3.2.2. Relative Motion Calculation

As shown in Figure 7, the binocular initial pose estimation can be divided into the following steps.

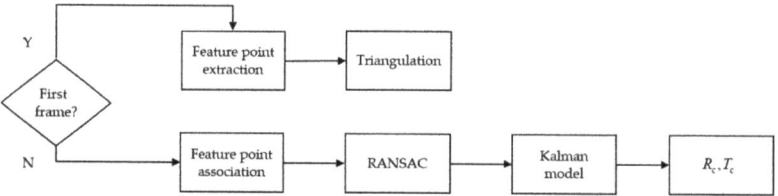

Figure 7. Flow chart of the initial pose acquisition phase.

(a) For the first frame, the feature point extraction is performed in the left and right images. A set of feature point pairs is obtained through brute force matching and the epipolar constraint criterion between the left and right cameras. Its three-dimensional coordinates in the left camera coordinate system are calculated through the principle of triangulation.

(b) For a non-first frame, the number of feature points will decrease if the optical flow tracking time is too long. If the number of feature points tracked in the current frame is less than threshold T_N (since perspective-n-point problems require at least 3 sets of points, the threshold should be greater than 3, which was set as 6 in this manuscript), then this frame adopts the method of the first frame to add new feature points. After obtaining the 3D set in the left camera coordinate system of the previous frame and the 2D feature point set of the left image in this frame, the rotation matrix and translation vector are solved quickly based on the random sample consensus (RANSAC) [27] method.

(c) Since the surface of a non-cooperative spacecraft has multilayer reflective materials, the texture information is relatively lacking, which leads to the appearance of outliers in the binocular measurement process. A Kalman filter model [28] was introduced to eliminate the outliers and ensure the stability of the system in this paper.

Assuming that the acceleration of the tracking spacecraft relative to the target spacecraft is constant, the geometric kinematics between the two spacecrafts is modeled. The state and observation of the system can be represented as:

$$X_k = A(X_{k-1}) + W_{k-1} \tag{13}$$

$$Z_k = H(X_k) + V_k \tag{14}$$

The state vector and observation vector are defined as $X_k = [x, y, z, \theta_x, \theta_y, \theta_z, \dot{x}, \dot{y}, \dot{z}, \dot{\theta}_x, \dot{\theta}_y, \dot{\theta}_z]^T$ and $Z_k = [\hat{x}, \hat{y}, \hat{z}, \hat{\theta}_x, \hat{\theta}_y, \hat{\theta}_z]^T$. W_{k-1} and V_k are the noise vector of the system and the noise vector in the observation process, respectively. The state transition matrix and observation matrix of the system can be defined as:

$$A_k = \begin{bmatrix} I_{3\times3} & O_{3\times3} & \Delta t \cdot I_{3\times3} & O_{3\times3} \\ O_{3\times3} & I_{3\times3} & O_{3\times3} & \Delta t \cdot I_{3\times3} \\ O_{3\times3} & O_{3\times3} & I_{3\times3} & O_{3\times3} \\ O_{3\times3} & O_{3\times3} & O_{3\times3} & I_{3\times3} \end{bmatrix}, H_k = \begin{bmatrix} I_{3\times3} & O_{3\times3} & O_{3\times3} & O_{3\times3} \\ O_{3\times3} & I_{3\times3} & O_{3\times3} & O_{3\times3} \end{bmatrix} \tag{15}$$

The predicted value of the filter is employed as the rotation and translation vector of the current frame to ensure that the target can still be tracked stably during the initial visual pose acquisition when abnormal values appear in the measurement process.

3.3. Point Cloud Tracking and Pose Optimization

3.3.1. Initial Value Calculation in the Point Cloud Coordinate System

In Section 3.2.2, the relative motion relationship between the front and back moments in the left camera coordinate system is acquired. We can obtain the motion relationship in the point cloud coordinate system. Assume that R_c, T_c is the rotation matrix and translation vector in the left camera coordinate system and R_{cp}, T_{cp} is the rotation matrix and translation vector from the camera coordinate system to the point cloud coordinate system obtained in the joint calibration. For a point set M in space, the following motion holds at time t and $t + 1$:

$$M_c^{t+1} = R_c M_c^t + T_c \tag{16}$$

$$M_p^t = R_{cp} M_c^t + T_{cp} \tag{17}$$

$$M_p^{t+1} = R_{cp} M_c^{t+1} + T_{cp} \tag{18}$$

Combined with the above formula, we obtain:

$$M_p^{t+1} = R_{cp}R_c R_{cp}^{-1} M_p^t + R_{cp}(T_c - R_c R_{cp}^{-1} T_{cp}) \qquad (19)$$

$$\begin{cases} R_p = R_{cp} R_c R_{cp}^{-1} \\ T_p = R_{cp}(T_c - R_c R_{cp}^{-1} T_{cp}) \end{cases} \qquad (20)$$

where R_p, T_p are the rotation matrix and translation vector in the point cloud coordinate system to be solved.

3.3.2. Frame-to Frame Point Cloud Registration

The number of point clouds collected by the TOF camera is large at close range, which affects the registration speed of the point clouds at the front and back points. It is necessary to downsample the point clouds. The ICP algorithm is usually used to solve the transformation relationship between two sets of point clouds. The ICP algorithm has the disadvantage of being easily trapped in a local minimum. It is sensitive to the initial pose guess, otherwise the algorithm cannot converge on the correct result.

The standard ICP registration algorithm was used in this study. The estimated pose in stereo vision is regarded as the initial value input to improve the accuracy of the initial pose. Therefore, the initial guess of the pose variation between two datasets is provided by the binocular-based method. The feature points in stereo vision are mapped to the point cloud data and clustered to obtain several SURF-3D key points for the point cloud registration. As shown in Figure 8, the small, white squares represent the key points of SURF-3D after K-means clustering.

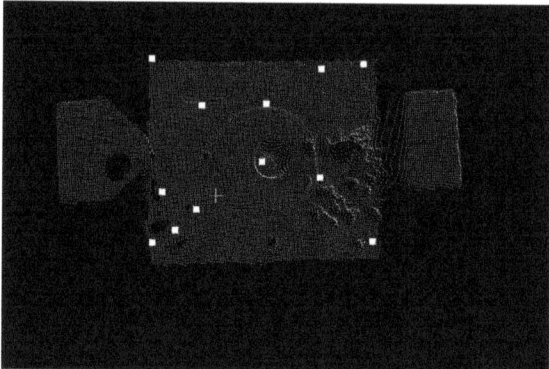

Figure 8. Schematic Pdiagram of the point cloud with the SURF-3D key points.

The steps of the frame-to-frame point cloud registration algorithm are shown in Figure 9.

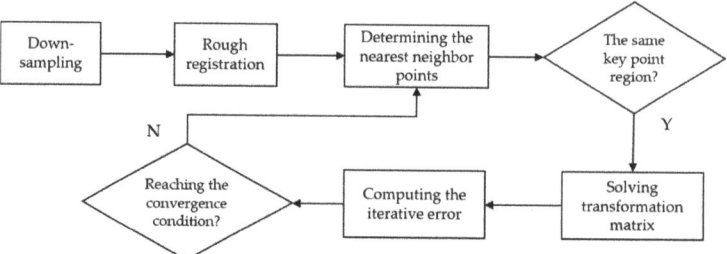

Figure 9. Flow chart of the frame-to-frame point cloud registration algorithm.

(a) Down-sampling the source point cloud using a voxel filter;
(b) Using the initial pose obtained by vision to perform the initial transformation on the source point cloud and accomplishing a rough registration of the point cloud;
(c) The K-dimensional tree is used to accelerate the search for point pairs between the source point cloud and the target point cloud when using the nearest neighbor to search for the corresponding points;
(d) It is judged whether the points belong to the same SURF-3D key point area when using the minimum value of the Euclidean distance to determine a point pair; the false matching phenomenon is rejected.
(e) The convergence condition is that the sum of the distance between the matched points is less than a given threshold or greater than the preset maximum number of iterations.

3.3.3. Pose Determination

Because of the necessity to provide a reference to the navigation system, according to our previous work, the docking ring of non-cooperative spacecraft was identified in the visual images to complete the absolute pose of the first frame. The reference absolute pose has five degrees of freedom. The detection results of the docking ring are shown in Figure 10.

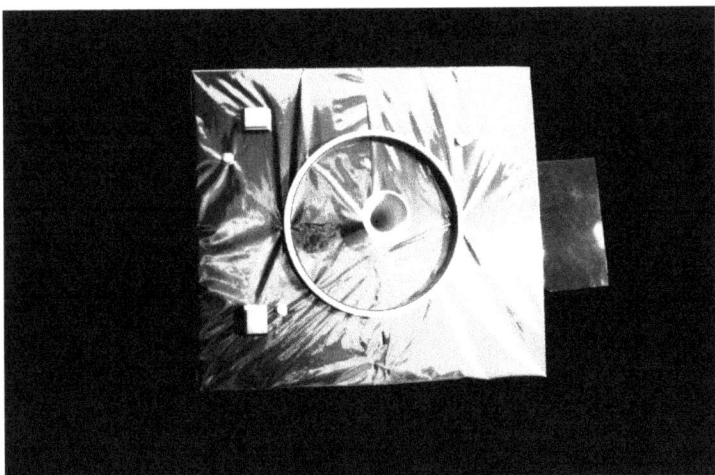

Figure 10. The image of the docking ring detection.

The pose transformation matrix of the current i-th frame can be represented as:

$$H_i^w = \begin{bmatrix} R_i^w & T_i^w \\ 0 & 1 \end{bmatrix} \tag{21}$$

where R_i^w and T_i^w represent the rotation matrix and the translation vector of the current frame, respectively. For the non-first frame, the pose transformation matrix of the j-th frame can be expressed as:

$$H_j^w = H_j^i H_i^{i-1} \cdots H_1^w \tag{22}$$

where H_j^i represents the transformation matrix from the i-th frame point cloud to the j-th frame point cloud. A pose correction method based on key frames is proposed in this study for the purpose of eliminating the cumulative error in the process of point cloud registration and tracking. The key frames selection criteria is as follows:

(a) The frame count difference between the current frame and the previous key frame is greater than the threshold T_c (it was set to 15 in this text), and the current frame is added to the key frame set;

(b) The relative movement distance between the current frame and the previous key frame is greater than the threshold T_d (it was set to 100 mm in this text) or the relative angle greater than the threshold T_a (it was set to 50 degree in this text); then, the current frame is added to the key frame set.

Inspired by the work in Reference [29], the pose objective function to be optimized in the key frame set can be defined as:

$$\arg\min \sum_{i,j} \frac{1}{3} \|E - (R_j^w)^{-1}(R_j^i R_i^w)\|_F^2 + \|T_j^w - T_i^w - R_i^w T_j^i\|_2^2 \tag{23}$$

where R_j^i, T_j^i represent the rotation matrix and translation vector from the i-th frame point cloud to the j-th frame point cloud, respectively. $\|\cdot\|_F$ represents the Frobenius norm of the matrix. The abovementioned nonlinear optimization problem is solved based on the Ceres library [30] in its specific implementation.

4. Experiment and Analysis
4.1. Numerical Simulation

In the simulation setting, the image resolution of the left and right cameras was 2048 × 2048 pixels, the rotation matrix between the two cameras was $R_c = I_{3\times 3}$, and the baseline, B, was 100 mm. The TOF resolution was 640 × 480 pixels, the rotation matrix between the left camera and TOF camera was $R_{cp} = I_{3\times 3}$, and the distance from left camera to the TOF camera was 50 mm.

The target rotated 180 degrees around the rolling axis in this simulation experiment. A total of 90 frames were collected. The angles of the X-, Y, and Z-axis are described as the yaw, pitch, and roll angle below, respectively. Figure 11a,b present the X-, Y-, and Z-axis position and Euler angle curve results during the low-speed rotation process.

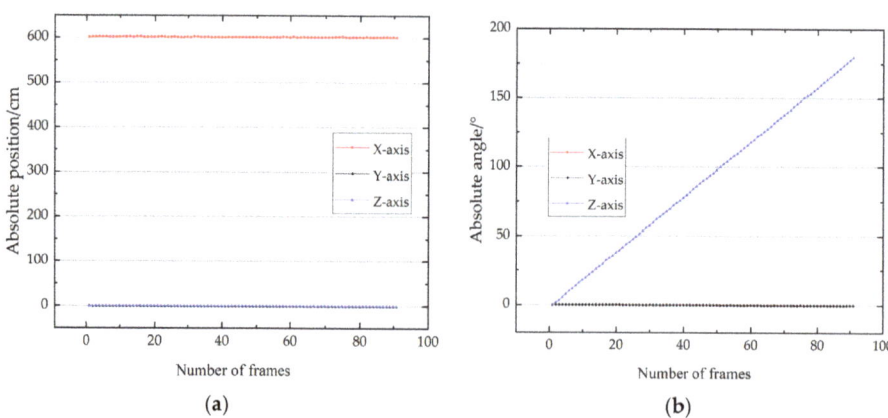

Figure 11. Pose estimation experimental results: (a) absolute position; (b) absolute angle.

It can be concluded that the absolute transformation also represents the position error curve. The maximum error of the X-, Y-, and Z-axis positions was 0.27, 0.88, and 1.94 mm. There is no doubt that the trend of the roll angle motion was consistent with the actual situation. The maximum error of the X-, Y-, and Z-axis angles was 0.32°, 0.31°, and 0.56°.

4.2. Semi-Physical Experiments
4.2.1. Ground Verification System

A ground experiment system, as shown in the figure below, was built in order to verify the feasibility of the method proposed in this paper. The whole system included two grayscale cameras (LUCID TRI032S-MC) and one TOF camera (LUCID HLT003S-001). The grayscale camera and TOF camera parameters are shown in Table 1. The non-cooperative satellite model was a 0.5 × 0.5 m cube model. The light in the space was handled through a solar simulator.

Table 1. Camera parameters.

	Grayscale Camera	TOF Camera
Sensor size	2048 × 1536 (pixel)	640 × 480 (pixel)
Pixel size	3.45 µm	10 µm
Field of view	41 × 31 (°)	69 × 51 (°)
Focal length	12 mm	8 mm

The rotational movement of the target was controlled by the ABB robotic arm. The ground truth was acquired by setting the dynamic data in advance and then driving the robotic arm to complete the corresponding movement through the program in the manipulator base coordinate system. The most important thing is that the frequency of the robotic arm movement and camera system acquisition was consistent. The accuracy of the ground truth was verified by an electronic total station. The premise of this scheme is that some reflection plates need to be artificially set on the satellite model. The circular calibration board was used to acquire the conversion relationship from the total station system to the camera system. In fact, the ground truth needs to be obtained every time through the above calibration scheme. However, due to the continuous motion, the preset dynamics data are regarded as the ground truth. The diagram is shown in Figure 12. Figure 13 illustrates the ground verification system.

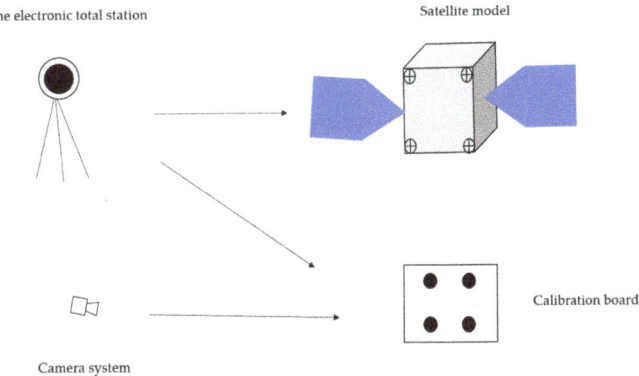

Figure 12. Schematic diagram to verify the accuracy of the ground truth.

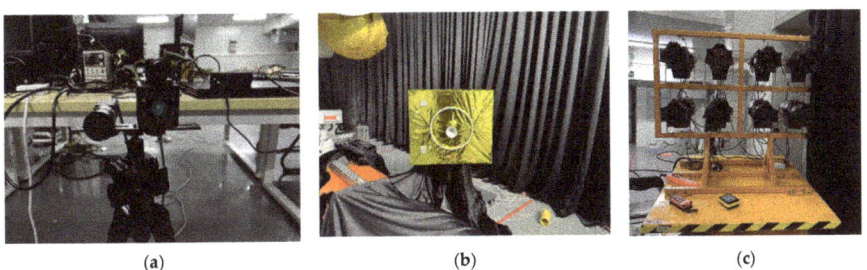

Figure 13. Schematic diagram of the ground verification system: (**a**) composite measurement system; (**b**) diagram of the non-cooperative target; (**c**) solar illumination simulator.

Then, the calibration work mentioned above was carried out. The internal parameters of the left camera and right camera were:

$$R_{kl} = \begin{bmatrix} 1761.482 & 0 & 1030.910 \\ 0 & 1758.598 & 781.315 \\ 0 & 0 & 1 \end{bmatrix}, R_{kr} = \begin{bmatrix} 1808.867 & 0 & 1027.288 \\ 0 & 1807.775 & 773.017 \\ 0 & 0 & 1 \end{bmatrix} \quad (24)$$

The calibration results of the external parameters were:

$$R_c = \begin{bmatrix} 0.9988 & -0.0004 & -0.0481 \\ 0.002 & 0.999 & -0.0042 \\ 0.0481 & 0.0042 & 0.9988 \end{bmatrix}, T_c = [-72.134, -61.887, -19.996]^T \text{ (mm)} \quad (25)$$

According to the joint calibration algorithm proposed in this paper, the calibration results from the camera coordinate system to the point cloud coordinate system were:

$$R_{cp} = \begin{bmatrix} 0.9999 & 0.0014 & -0.0011 \\ -0.0014 & 0.9999 & 0.0162 \\ 0.0011 & -0.0162 & 0.9999 \end{bmatrix}, T_{cp} = [-3.6327, -49.2564, -1.0740]^T \text{ (mm)} \quad (26)$$

The joint calibration results were verified by a reprojection experiment. As shown in Table 2, the reprojection error of the circular calibration board was much smaller than that of the checkerboard.

Table 2. Reprojection errors of the two calibration boards.

	Circular Board	Checkerboard
X-axis	0.95 pixel	1.78 pixels
Y-axis	0.78 pixel	1.71 pixels

4.2.2. Pose Measurement Experiment

The target was a specific distance from the camera system. The camera system was kept still, and the target rotated 360° under the control of the robotic arm to simulate the tumbling state of the non-cooperative target in the space environment. Several frames of visual images and point cloud were captured. Figure 14 shows the left and right camera images and the point cloud data in a specific state.

(a)

(b)
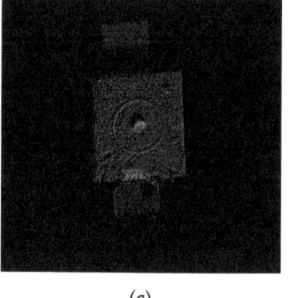
(c)

Figure 14. Left and right camera images and TOF camera data: (a) left image; (b) right image; (c) point cloud.

Rotation Experiment at Low Speed

The target was set to a low-speed rotation rate (approximately 6 degrees per frame) in this experiment. Figure 15a,b present the X-, Y-, and Z-axis position and Euler angle curve results during the low-speed rotation process. Figure 15c,d show the X-, Y-, and Z-axis position and Euler angle error results during the rotation process.

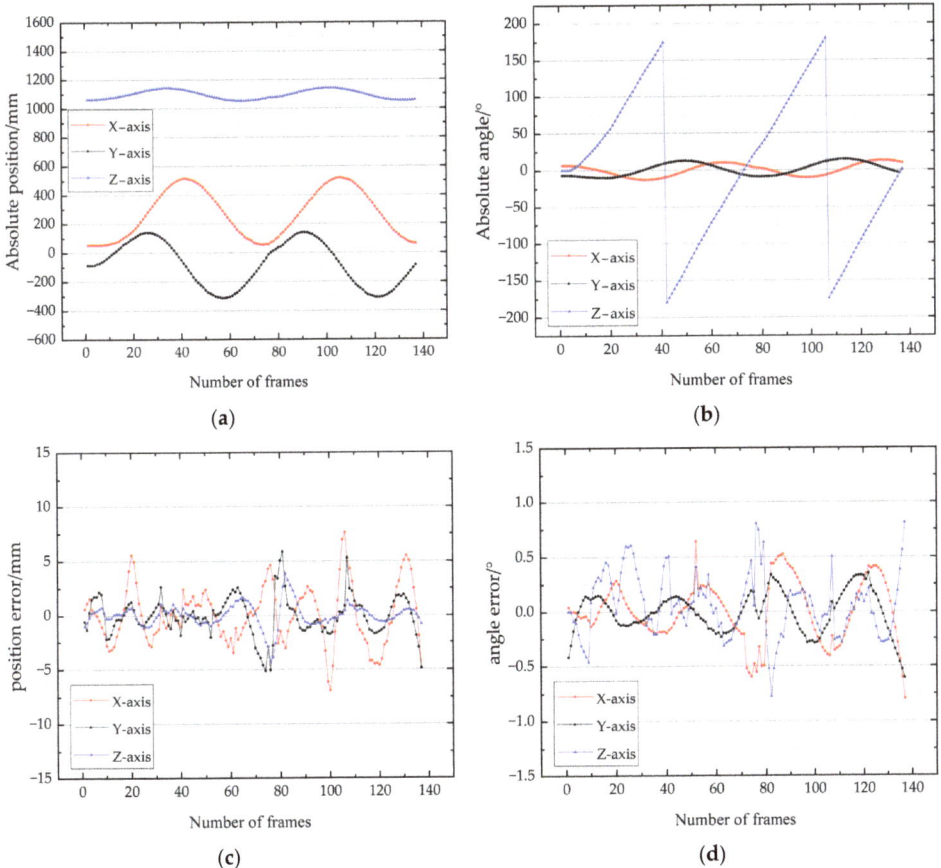

Figure 15. Pose estimation experimental results at low speed: (**a**) absolute position; (**b**) absolute angle; (**c**) error of position; (**d**) error of angle.

Since the measurement coordinate system of the camera system was not parallel to the coordinate system of the robot arm, the X-axis and Y-axis rotated with it simultaneously when the target rolled around the Z-axis. The X-axis and Y-axis positions had a large amount of movement. Both the pitch and yaw angles had nutation of approximately 20 degrees. The maximum error and average error of the X, Y, and Z three-axis position of the proposed method was 7.6, 5.9, and 4.4 mm and 2.1, 1.4, and 0.76 mm, respectively. The maximum error and average error of the X, Y, and Z three-axis angle of the proposed method were 0.81°, 0.61°, and 0.81° and 0.22°, 0.16°, and 0.23°, respectively.

Figure 16 illustrates the X, Y, and Z three-axis position and Euler angle error results obtained by the proposed method, the stereo vision method, the traditional fast point feature histograms (FPFH) [31], and the ICP algorithm throughout the low-speed process.

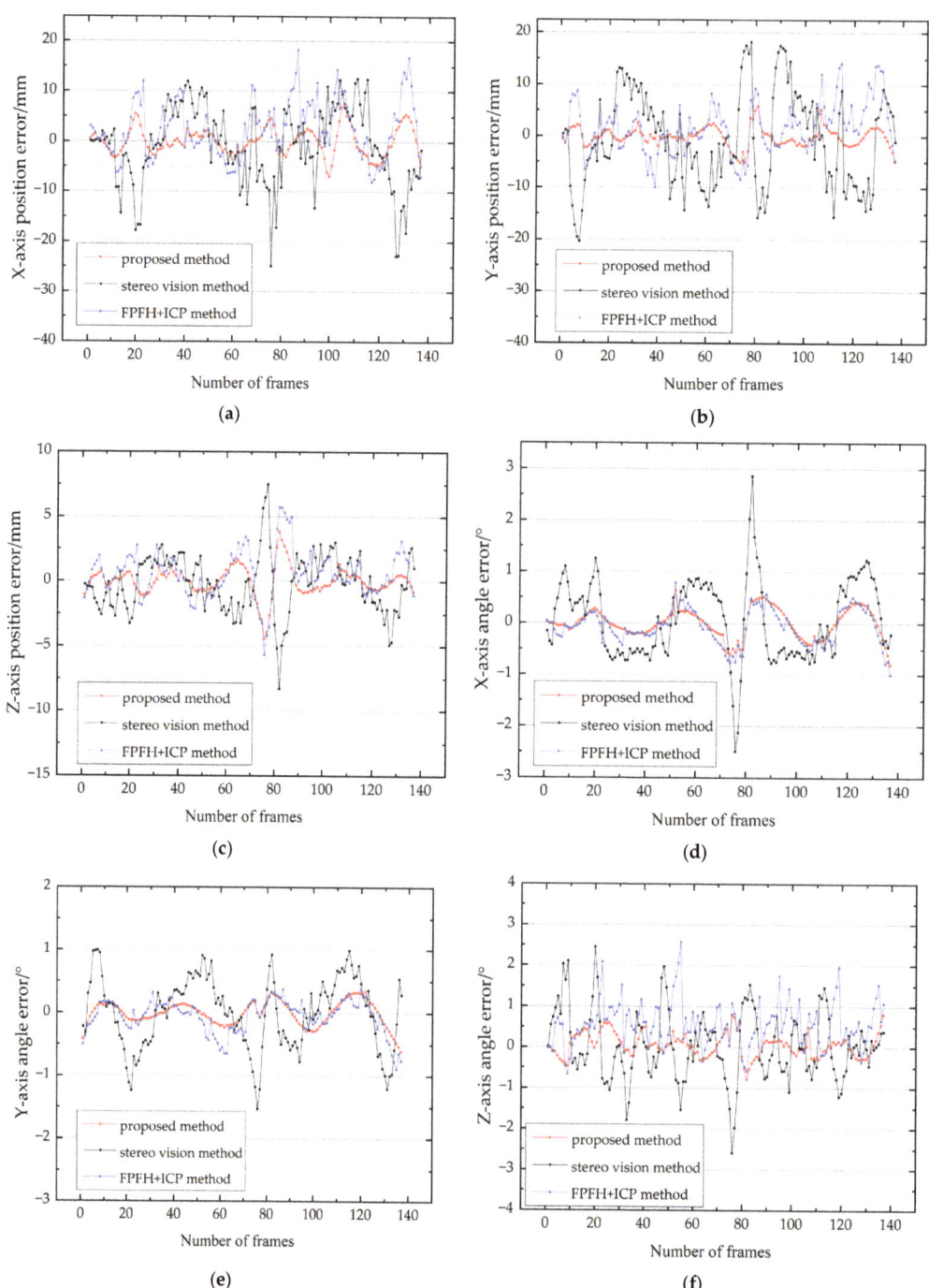

Figure 16. Comparison of the error results of the different methods at low speed: (**a**) X-axis position error; (**b**) Y-axis position error; (**c**) Z-axis position error; (**d**) X-axis angle error; (**e**) Y-axis angle error; (**f**) Z-axis angle error.

As shown in Figure 16, it can intuitively be seen that the pose result obtained from the stereo vision method had a larger error. The maximum error and average error of the X, Y, and Z three-axis position of the stereo vision method was 24.9, 20.4, and 8.3 mm and 6.2, 7.9, and 1.8 mm, respectively. The maximum error and average error of the X, Y, and Z three-axis angle of the stereo vision method was 2.85°, 1.53°, and 2.59° and 0.66°, 0.47°, and 0.67°, respectively.

The traditional FPFH algorithm can also obtain the initial value, mainly by applying the fast point feature histogram to extract the key features, employing the initial sampling consistency registration algorithm for rough registration. The ICP fine registration and pose correction were conducted in this comparison experiment. The maximum error and average error of the X, Y, and Z three-axis position of the FPFH method was 18.3, 14.2, and 5.7 mm and 4.8, 4.0, and 1.4 mm, respectively. The maximum error and average error of the X, Y, and Z three-axis angle of the FPFH method were 0.99°, 0.88°, and 2.57° and 0.28°, 0.20°, and 0.64°, respectively.

Rotation Experiment at High Speed

The kinetic data are consistent with those in Section Rotation Experiment at Low Speed. The experimental conditions were set at a relatively high-speed rotation (approximately 12 degrees per frame). Figure 17a,b show the X-, Y-, and Z-axis position and Euler angle curve results during the high-speed rotation process. Figure 17c,d illustrate the X-, Y-, and Z-axis position and Euler angle error results during the rotation process.

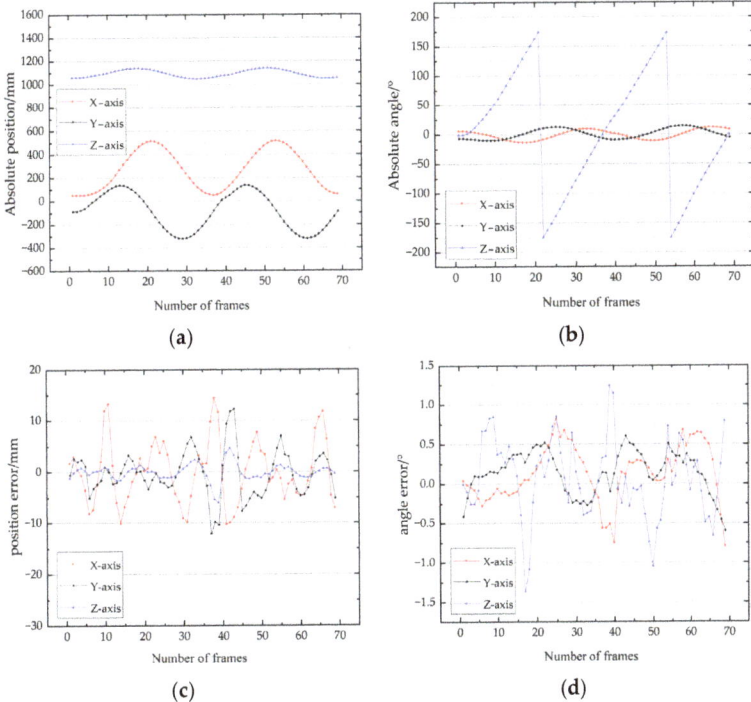

Figure 17. Pose estimation experimental results at high speed: (**a**) absolute position; (**b**) absolute angle; (**c**) error of position; (**d**) error of angle.

The method proposed in this paper can cope with the high-speed rotation condition from the data in the figure above. The target could be tracked stably in the high-speed experiment. The maximum error and average error of the X, Y, and Z three-axis position of the proposed method were 14.4, 12.3, and 5.9 mm and 4.9, 3.5, and 1.08 mm, respectively. The maximum error and average error of the X, Y, and Z three-axis angle of the proposed method were 0.82°, 0.60°, and 1.36° and 0.29°, 0.26°, and 0.39°, respectively. Contrasted with the working conditions in Section Rotation Experiment at Low Speed, the position and angle errors increased, which still indicates that the task of capturing non-cooperative spacecrafts can be completed.

Figure 18 represents the X, Y, and Z three-axis position and Euler angle error results obtained by the proposed method, the stereo vision algorithm, and the traditional FPFH + ICP algorithm in the high-speed process.

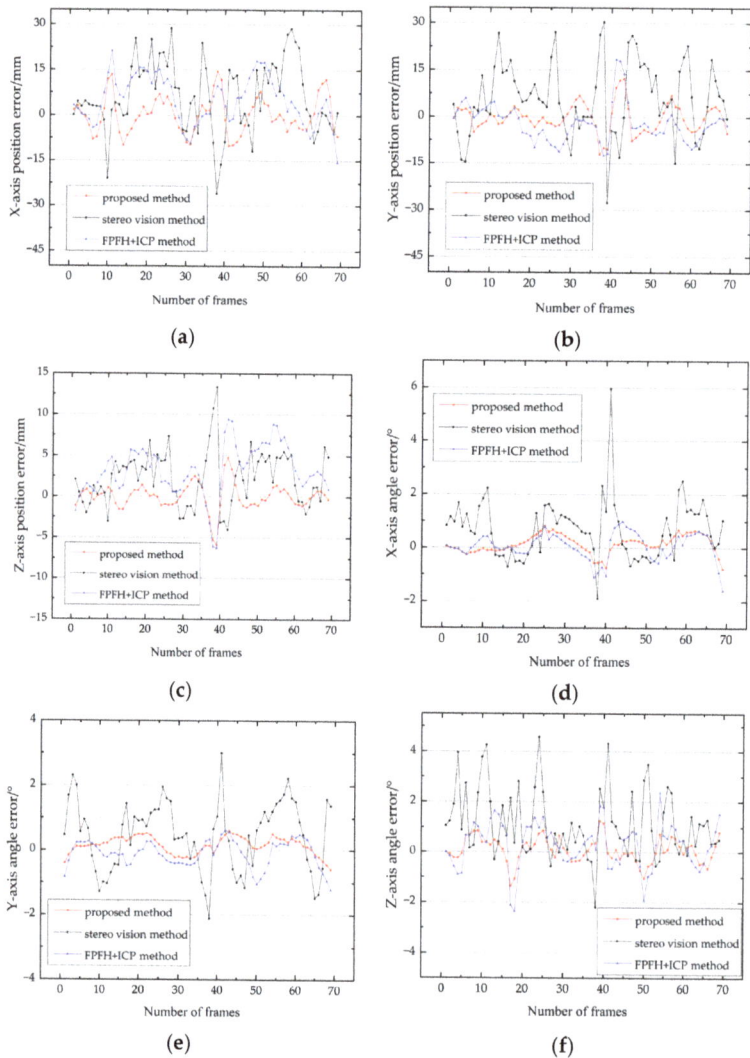

Figure 18. Comparison of the error results of the different methods at high speed: (**a**) X-axis position error; (**b**) Y-axis position error; (**c**) Z-axis position error; (**d**) X-axis angle error; (**e**) Y-axis angle error; (**f**) Z-axis angle error.

The maximum error and average error of the X, Y, and Z three-axis position of the stereo vision method were 28.7, 30.4, 13.4 mm and 10.4, 10.4, and 3.2 mm, respectively. The maximum error and average error of the X, Y, and Z three-axis angle of the stereo vision method were 5.9°, 3.0°, and 4.6° and 0.96°, 1.01°, and 1.3°, respectively.

The error trend of the FPFH method in the high-speed rotation was coincident with the proposed method. The maximum error and average error of the X, Y, and Z three-axis position of the FPFH method were 21.3, 18.5, and 9.41 mm and 7.4, 4.8, and 3.7 mm, respectively. The maximum error and average error of the X, Y, and Z three-axis angle of the FPFH method were 1.6°, 1.21°, and 2.36° and 0.39°, 0.34°, and 0.77°, respectively.

Key Frames Threshold Selection Experiment

The selection of the key frames threshold had a great influence on eliminating the cumulative error in the process of pose measurement. Figure 19 shows the pose errors resulting from the different thresholds.

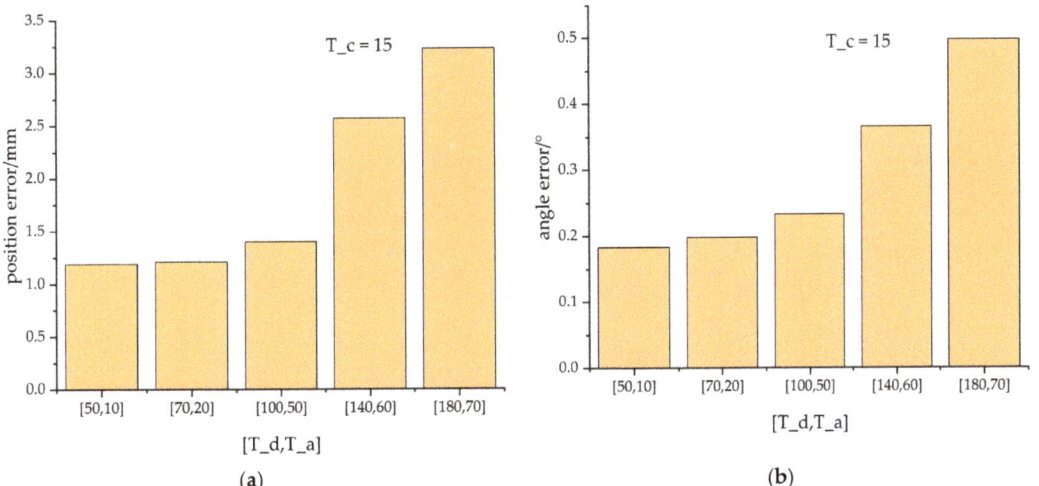

Figure 19. The pose errors resulting from different thresholds: (**a**) position error; (**b**) angle error.

As can be seen from the figure, the selection of the key frames threshold in this manuscript was appropriate. Small thresholds will lead to more memory consumption and increase the amount of computation in the pose measurement system. The suitable thresholds T_d and T_a can prevent the error drift phenomenon in the rotation process. The threshold T_c updates the set of key frames when the relative position and angle do not meet the conditions for a long time. From the perspective of memory and computation, more ground experiments are required to select the optimal threshold values in practical application.

Calculation Time Comparison of Initial Value Acquisition Methods

For a point cloud with N points and K-neighbors for each point, the computational complexity of the FPFH method is O(NK). However, the stereo vision method is O(M). M is the number of feature points. For the improved ICP point cloud registration step, the complexity of the algorithm is O(N). As shown in the figure below, the calculation times of the initial pose obtained through the proposed method and FPFH method were compared. All mentioned methods were implemented on a PC (I7-8700 at 3.2 GHz, 16 GB RAM) with Visual Studio 2019. The programming language was C++. The OPENCV library, Point Cloud Library (PCL), and Ceres optimization library were used in this study. As shown in Figure 20, the calculation time of different initial value acquisition methods are compared.

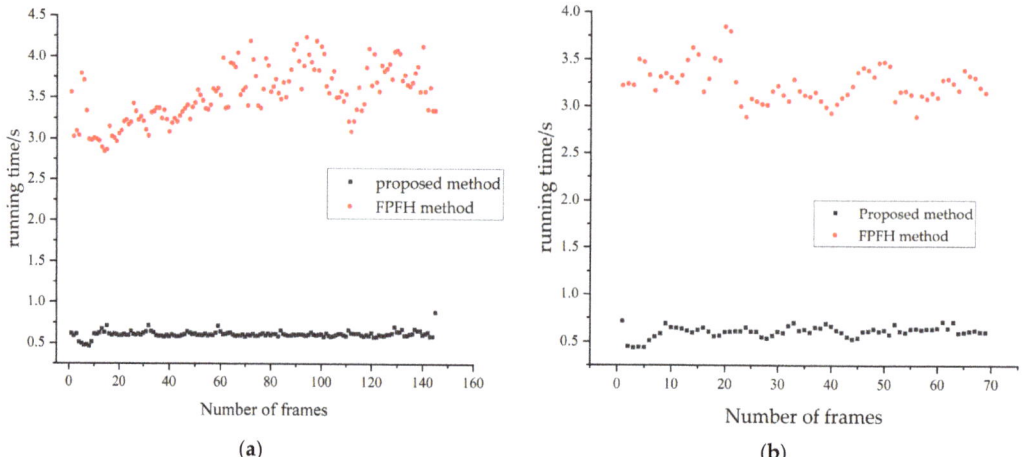

Figure 20. Comparison of the computational time between the proposed method and the FPFH method in rotation experiments: (**a**) low-speed experiment; (**b**) high-speed experiment.

The running time of the initial pose obtained using the proposed method fluctuated around 0.6 s. However, the value acquired by the FPFH method was within 2.5 to 4.5 s. There was an improvement in the initial pose acquisition time when using the proposed method.

In summary, the relative pose error estimated in this paper was the smallest in comparison to similar experiments. The method in this paper displayed less running time, which demonstrates that it is more conducive to the application of actual control.

5. Conclusions

A method for the pose measurement of non-cooperative spacecrafts based on binocular and TOF camera collaboration was proposed. Firstly, a joint calibration method between the binocular camera and TOF camera based on a circular calibration board was conducted. The reprojection error demonstrated the effectiveness of the proposed calibration method. Then, the initial pose method was mainly divided into feature extraction, data association, and Kalman model suppression. The frame-to-frame ICP registration and pose correction based on key frames were carried out during the point cloud tracking and pose optimization. The pose results of the proposed method, stereo vision method, and the traditional FPFH method were compared in the ground verification experiment. The experimental results show that the position error of the proposed method was within 1 cm, and the angle error was within 1 degree in a low-speed rotation process. The position and angle error were within 1.5 cm and 1.4 degrees during the high-speed rotation conditions, respectively. The proposed method had a certain accuracy and robustness when chasing the target satellite, especially for a satellite with an unknown structure. In further studies, the relative orbital and attitude dynamics should be considered in the Kalman filtering process, and the proposed method should be implemented on the embedded platform to verify the real-time performance.

Author Contributions: Conceptualization, L.H. and H.D.; methodology, L.H. and D.S.; Validation, L.H., D.S. and A.S.; formal analysis, S.Z.; investigation, H.P.; resources, H.P.; data curation, L.H., D.S. and A.S.; writing—original draft preparation, L.H.; writing—review and editing, D.S. and H.D.; supervision, S.Z.; funding acquisition, H.P. All authors have read and agreed to the published version of the manuscript.

Funding: This research was funded by the Preliminary Research Foundation of Equipment, grant number: 3050404030, and the Innovation Program CX-325 of the Shanghai Institute of Technical Physics.

Institutional Review Board Statement: Not applicable.

Informed Consent Statement: Not applicable.

Data Availability Statement: Not applicable.

Conflicts of Interest: The authors declare no conflict of interest.

References

1. Takeichi, N.; Tachibana, N. A tethered plate satellite as a sweeper of small space debris. *Acta Astronaut.* **2021**, *189*, 429–436. [CrossRef]
2. Muntoni, G.; Montisci, G.; Pisanu, T.; Andronico, P.; Valente, G. Crowded Space: A Review on Radar Measurements for Space Debris Monitoring and Tracking. *Appl. Sci.* **2021**, *11*, 1364. [CrossRef]
3. Razzaghi, P.; Al Khatib, E.; Bakhtiari, S.; Hurmuzlu, Y. Real time control of tethered satellite systems to de-orbit space debris. *Aerosp. Sci. Technol.* **2021**, *109*, 106379. [CrossRef]
4. Mark, C.P.; Kamath, S. Review of active space debris removal methods. *Space Policy* **2019**, *47*, 194–206. [CrossRef]
5. Liu, J.; Tong, Y.; Liu, Y.; Liu, Y. Development of a novel end-effector for an on-orbit robotic refueling mission. *IEEE Access.* **2020**, *8*, 17762–17778. [CrossRef]
6. Li, W.; Cheng, D.; Liu, X.; Wang, Y.; Shi, W.; Tang, Z.; Gao, F.; Zeng, F.; Chai, H.; Luo, W.; et al. On-orbit service (OOS) of spacecraft: A review of engineering developments. *Prog. Aerosp. Sci.* **2019**, *108*, 32–120. [CrossRef]
7. Moghaddam, B.M.; Chhabra, R. On the guidance, navigation and control of in-orbit space robotic missions: A survey and prospective vision. *Acta Astronaut.* **2021**, *184*, 70–100.
8. Luu, M.A.; Hastings, D.E. On-Orbit Servicing System Architectures for Proliferated Low-Earth-Orbit Constellations. *J. Spacecr. Rocket.* **2022**, *59*, 1946–1965. [CrossRef]
9. Oestreich, C.; Espinoza, A.T.; Todd, J.; Albee, K.; Linares, R. On-Orbit Inspection of an Unknown, Tumbling Target Using NASA's Astrobee Robotic Free-Flyers. In Proceedings of the IEEE/CVF Conference on Computer Vision and Pattern Recognition, Nashville, TN, USA, 20–25 June 2021; pp. 2039–2047.
10. Huang, X.; Li, M.; Wang, X.; Hu, J.; Zhao, Y.; Guo, M.; Xu, C.; Liu, W.; Wang, Y.; Hao, C.; et al. The Tianwen-1 guidance, navigation, and control for Mars entry, descent, and landing. *Space Sci.Technol.* **2021**, *2021*, 9846185.
11. Zhao, G.; Xu, S.; Bo, Y. LiDAR-based non-cooperative tumbling spacecraft pose tracking by fusing depth maps and point clouds. *Sensors* **2018**, *18*, 3432. [CrossRef]
12. Hu, Q.; Jiang, C. Relative Stereovision-Based Navigation for Noncooperative Spacecraft via Feature Extraction. *IEEE/ASME Trans. Mechatron.* **2022**, *27*, 2942–2952. [CrossRef]
13. Zhang, L.; Zhu, F.; Hao, Y.; Pan, W. Rectangular-structure-based pose estimation method for non-cooperative rendezvous. *Appl. Opt.* **2018**, *57*, 6164–6173. [CrossRef] [PubMed]
14. Peng, J.; Xu, W.; Liang, B.; Wu, A. Virtual stereovision pose measurement of noncooperative space targets for a dual-arm space robot. *IEEE Trans. Instrum. Meas.* **2019**, *69*, 76–88. [CrossRef]
15. Li, Y.; Jia, Y. Stereovision-based Relative Motion Estimation Between Non-cooperative spacecraft. In Proceedings of the 2019 Chinese Control Conference (CCC), Guangzhou, China, 27–30 July 2019; pp. 4196–4201.
16. Liu, L.; Zhao, G.; Bo, Y. Point cloud based relative pose estimation of a satellite in close range. *Sensors* **2016**, *16*, 824. [CrossRef] [PubMed]
17. Opromolla, R.; Fasano, G.; Rufino, G.; Grassi, M. Pose estimation for spacecraft relative navigation using model-based algorithms. *IEEE Trans. Aerosp. Electron. Syst.* **2017**, *53*, 431–447. [CrossRef]
18. Guo, W.; Hu, W.; Liu, C.; Lu, T. Pose initialization of uncooperative spacecraft by template matching with sparse point cloud. *J. Guid. Control Dyn.* **2021**, *44*, 1707–1720. [CrossRef]
19. Terui, F.; Kamimura, H.; Nishida, S. Motion estimation to a failed satellite on orbit using stereo vision and 3D model matching. In Proceedings of the 2006 9th International Conference on Controll, Automation, Robotics and Vision, Singapore, 5–8 December 2006; pp. 1–8.
20. Peng, J.; Xu, W.; Liang, B.; Wu, A. Pose measurement and motion estimation of space non-cooperative targets based on laser radar and stereo-vision fusion. *IEEE Sens. J.* **2018**, *19*, 3008–3019. [CrossRef]
21. Guo, P.; Zhang, Y.; Hu, Q. Pose Measurement of Non-cooperative Spacecraft by Sensors Fusion. In Proceedings of the 2022 41st Chinese Control Conference (CCC), Hefei, China, 25–27 July 2022; pp. 3426–3431.
22. Liu, Z.; Liu, H.; Zhu, Z.; Song, J. Relative pose estimation of uncooperative spacecraft using 2D–3D line correspondences. *Appl. Opt.* **2021**, *60*, 6479–6486. [CrossRef]
23. Su, Y.; Zhang, Z.; Wang, Y.; Yuan, M. Accurate Pose Tracking for Uncooperative Targets via Data Fusion of Laser Scanner and Optical Camera. *J. Astronaut. Sci.* **2022**, 1–19.
24. Sun, D.; Hu, L.; Duan, H.; Pei, H. Relative Pose Estimation of Non-Cooperative Space Targets Using a TOF Camera. *Remote Sens.* **2022**, *14*, 6100. [CrossRef]
25. Vidmar, A.; Brilly, M.; Sapač, K.; Kryžanowski, A. Efficient Calibration of a Conceptual Hydrological Model Based on the Enhanced Gauss–Levenberg–Marquardt Procedure. *Appl. Sci.* **2020**, *10*, 3841. [CrossRef]
26. Zhang, G.; Qin, D.; Yang, J.; Yan, M.; Tang, H.; Bie, H.; Ma, L. UAV Low-Altitude Aerial Image Stitching Based on Semantic Segmentation and ORB Algorithm for Urban Traffic. *Remote Sens.* **2022**, *14*, 6013. [CrossRef]

27. Fischler, M.A.; Bolles, R.C. Random sample consensus: A paradigm for model fitting with applications to image analysis and automated cartography. *Commun. ACM* **1981**, *24*, 381–395. [CrossRef]
28. Urrea, C.; Agramonte, R. Kalman filter: Historical overview and review of its use in robotics 60 years after its creation. *J. Sensors* **2021**, *2021*, 9674015. [CrossRef]
29. Wang, Q.; Cai, G. Pose estimation of a fast tumbling space noncooperative target using the time-of-flight camera. *Proc. Inst. Mech. Eng. Part G J. Aerosp. Eng.* **2021**, *235*, 2529–2546. [CrossRef]
30. Agarwal, S.; Mierle, K. Ceres solver: Tutorial & reference. *Google Inc* **2012**, *2*, 8.
31. Rusu, R.B.; Blodow, N.; Beetz, M. Fast point feature histograms (FPFH) for 3D registration. In Proceedings of the 2009 IEEE International Conference on Robotics and Automation, Kobe, Japan, 12–17 May 2009; pp. 3212–3217.

Disclaimer/Publisher's Note: The statements, opinions and data contained in all publications are solely those of the individual author(s) and contributor(s) and not of MDPI and/or the editor(s). MDPI and/or the editor(s) disclaim responsibility for any injury to people or property resulting from any ideas, methods, instructions or products referred to in the content.

Article

Hybrid-Compliant System for Soft Capture of Uncooperative Space Debris

Maxime Hubert Delisle *,†, Olga-Orsalia Christidi-Loumpasefski †, Barış C. Yalçın, Xiao Li, Miguel Olivares-Mendez and Carol Martinez

Space Robotics Research Group, SnT Interdisciplinary Centre for Security, Reliability and Trust, University of Luxembourg, 6 Rue Richard Coudenhove-Kalergi, 1359 Luxembourg, Luxembourg; miguel.olivaresmendez@uni.lu (M.O.-M.)
* Correspondence: maxime.hubertdelisle@uni.lu
† These authors contributed equally to this work.

Featured Application: The proposed hybrid-compliant concept is meant to be part of a space debris capture system.

Abstract: Active debris removal (ADR) is positioned by space agencies as an in-orbit task of great importance for stabilizing the exponential growth of space debris. Most of the already developed capturing systems are designed for large specific cooperative satellites, which leads to expensive one-to-one solutions. This paper proposed a versatile hybrid-compliant mechanism to target a vast range of small uncooperative space debris in low Earth orbit (LEO), enabling a profitable one-to-many solution. The system is custom-built to fit into a CubeSat. It incorporates active (with linear actuators and impedance controller) and passive (with revolute joints) compliance to dissipate the impact energy, ensure sufficient contact time, and successfully help capture a broader range of space debris. A simulation study was conducted to evaluate and validate the necessity of integrating hybrid compliance into the ADR system. This study found the relationships among the debris mass, the system's stiffness, and the contact time and provided the required data for tuning the impedance controller (IC) gains. This study also demonstrated the importance of hybrid compliance to guarantee the safe and reliable capture of a broader range of space debris.

Keywords: space debris; active debris removal; impedance controller; in-orbit servicing; uncooperative satellites; gecko-inspired dry adhesive

1. Introduction

Since humankind initiated space activities more than 60 years ago, the number of in-orbit objects has increased [1]. More than 330 million debris objects not bigger than 1 cm are in orbit. The number of objects between 1 and 10 cm is close to 1 million, whereas there are around 36,500 debris objects greater than 10 cm [2]. The Kessler Syndrome states that the amount of space debris is growing exponentially [3], which leads to a crucial problem for ongoing and future space missions. Two approaches have been proposed to mitigate the space debris problem—active debris removal (ADR) and passive debris removal (PDR). However, PDR cannot achieve the desired stabilized number of debris in the foreseeable future. Even if space launches stop, the number of space debris would still increase due to future collisions. Therefore, ADR is required [4].

The problem with space debris is that most targets are not designed for removal. They are uncooperative for capturing [5] and they do not include specific grippers, handles, or markers to make capturing easier [6]. Additionally, each debris object has a unique geometry, velocity, and material [7]. The fact that they can be tumbling at hyper-velocity constitutes a crucial danger at any orbit [8]. Hence, capturing autonomously and harmlessly uncooperative objects demands reliability, robustness, and control at the impact, as the space environment and the crucial nature of the mission are demanding. These requirements give the capturing phase the most critical role in the mission.

Capturing mechanisms can interact differently with the debris; an Energy-Transfer Classification (ET-Class) was proposed in [9]. For instance, the *Impact Energy Dissipation (ET2)* class implies a capture with a decrease of energy at the first impact. In this class, the rigid and flexible capturing methods stand out as the most promising for their reliability. The rigid capturing operation was one of the first capturing methods tried for the realization of mechanical contact in space. However, this method requires, in some cases, extremely expensive motion control since any misalignment during the contact can push the debris far away [10,11], especially if the object is tumbling at high velocity [12]. Additionally, any rigid capturing mechanism must be lightweight and compatible with different space debris volumes [13]. As a result, the rigid capturing method is more applicable for cooperative targets that have proper docking ports [14].

In the literature, many rigid robotic structures are from single-arm to multiple-arms [15,16]. Multiple arms are controlled by more complex control algorithms, such as sliding mode control or adaptive control [17]. For single-arm rigid capturing methods, the classical PID control approach is enough to achieve position and velocity control of the end-effector [18]. Nowadays, reinforcement learning (RL)- [19,20], model predictive control (MPC)- [21], and $\mathcal{H}_\infty$-based [22] methods are also researched. Yet, the most crucial problem regarding rigid capturing methods remains the same, which is the difficulty of achieving robust mechanical interaction using rigid structures in space, since rigidness lacks appropriate impact energy dissipation in a frictionless environment. Therefore, both academic and industrial research are inclined to focus on flexible capturing methods rather than rigid capturing methods [23]. Regarding flexible capturing methods, shape memory alloys (SMA) and pneumatic capturing mechanisms are nowadays part of the most popular flexible mechanisms [24–26]. They can be categorized in the *ET2* category, as capturing mechanisms of this class decrease the impact energy of the debris at the very first contact, according to the ET-Class. For example, capturing mechanisms using SMA material can fully comply with the debris geometry. Moreover, the actuation of SMA does not demand high energy consumption [27,28]. The most sophisticated study accomplished in this field is MEDUSA. MEDUSA has flexible arms actuated by electrical inputs that can grasp nearly any object. When a simple electrical signal triggers the nitinol wires, the arms of MEDUSA begin to adapt their shape and grasp space debris [29]. Many detailed experiments showed the great robustness of the mechanical contact. However, despite promising on-ground facility results, these capturing methods have not been tested in space yet, making their performance fuzzy for on-site space applications.

In addition, more flexible mechanisms take advantage of the gecko-inspired dry adhesive to stick to the debris surface. However, they are, so far, either not suitable for small autonomous integration applications [30] or fitting a specific debris shape [31,32]. Moreover, a critical factor for a capturing system is its ability to absorb the first impact with the target; a strong and rigid impact can lead to mechanical failure and, thus, to mission failure or debris generation. To deal with this, researchers have integrated either passive [33] or active [34] compliance into their systems. However, to ensure adequate contact time with the debris for the adhesive to stick to the debris surface during the impact, passive compliance, although essential to dissipate the impact energy, is not enough; controlled active compliance of the interaction is required [35]. To the authors' knowledge, no such hybrid system, i.e., a system with both active and passive compliance, has yet been proposed.

Therefore, this paper proposes the following:

- A concept for an active debris removal capturing phase (Section 2);
- A hybrid-compliant system for the soft capture of space debris (Section 3);
- An impedance control design for the proposed hybrid-compliant system (Section 4).

The proposed flexible, versatile hybrid-compliant system of class *ET2* [9], custom-built to fit a CubeSat, is displayed in Figure 1. This new system targets a vast range of small debris, enabling a profitable one-to-many solution. In contrast to previous concepts, the mechanism's compliance is hybrid. It incorporates active (with linear actuators and impedance controller) and passive (thanks to revolute joints) compliance to dissipate the impact energy, allow adequate contact time, and successfully help capture a broader range of space debris.

Figure 1. Concept of a CubeSat-based system for capturing small debris.

Impedance control (IC) is an example of active interaction control, incorporating lumped parameters [36]. For a mechanism in contact with debris, IC can regulate the relationship between the mechanism's tip position and the impact force [37,38]. An essential part of IC is the proper tuning of its gains. By adjusting them regarding the mass of the debris to be captured, the capturing system can target a wider range of debris. In this paper, a simulation study was conducted. It presented the correlation between the debris mass, the ADR system's hybrid compliance, and the contact time, providing the required data for appropriate IC gains tuning. In addition, the necessity of hybrid compliance and the IC was validated.

This paper is organized as follows. Section 2 introduces the space debris capture problem with a brief on space environment statistics, focusing on LEOs to determine which shape is the most common and must be targeted first. Additionally, the section presents the proposed concept of operations (ConOps) for an ADR Capturing Phase. Section 3 presents the proposed hybrid-compliant system, and its integration into the proposed capturing phase is described. Section 4 introduces the impedance controller, a critical component of the proposed hybrid-compliant system. Finally, Section 5 presents the simulation study and discussion of the results, and Section 6 presents the conclusions and direction of future work.

2. Space Debris Capture

Despite the growing concern about space debris, no autonomous capturing system has been officially used yet. The required technologies can be quite diverse and by 2025 we will see the launch of the first autonomous chaser satellites by ClearSpace to remove an ESA-owned item from orbit (ClearSpace-1 mission [39,40]).

ADR missions depend a lot on the targeted debris. The most commonly studied solution is to design one capturing system for one specific debris (one-to-one solution). Currently, voluminous and well-known satellites are the ones aimed to be targeted first. However, although these satellites are one of the main threats to generating more space debris, it is only one side of the problem. The new mega-constellations of CubeSats coming in the next decade in LEO (around the 500–700 km orbits) will increase the number of decommissioned satellites remaining in orbit. As a result, the urge to tackle the small satellites in LEO is and will be real.

The capturing mechanism plays a key role in the success of an autonomous space debris removal mission, especially if it is designed to target a wide range of debris, as the one proposed in this paper. To that extent, to design such a system, it is of utmost importance to know about the variety of objects in LEO, obtain knowledge of that data, and determine what range of debris our mechanism should target first. These parameters will impact the design of an autonomous ADR system.

2.1. Debris Data

Space environment statistics is a new space debris topic addressing debris tracking. Due to the technological limitations of the surveillance networks, small-size debris is currently not trackable. In December 2022, more than 32,500 objects were regularly tracked by space surveillance networks. In contrast, more than 130 million objects starting from 1 mm in size are estimated to be in space orbit, based on statistical models [2]. The growing space debris issue in LEO creates the need for knowledge about those objects to design adequate debris removal systems. As ESA made available the catalogue of the tracked objects via the single-source DISCOS (Database and Information System Characterising Objects in Space) dataset [41], which is updated every few months, it is possible to analyze the LEO debris population. DISCOS plays a daily role in some of the ESA activities, such as collision avoidance, re-entry analyses, and for contingency support.

By analyzing the DISCOS dataset, a debris population of almost 20,000 objects with nearly 300 different shapes was found in LEO. For each object, the available features are their mass, shape (with size characteristics, when available), and information about their orbits (apogee, perigee). All these objects could potentially threaten any space mission. However, we prioritize the shapes more commonly found in LEO (found more than a hundred times) for designing our capturing mechanism. Additionally, as the focus is on small satellite removal, the targeted debris' size and mass are non-negligible factors. To that extent, we narrowed down the catalogue of objects in LEO to those lower than or equal to 100 kg.

The total number of objects found in LEO with the mentioned parameters was 4162. Among the 107 different specific shapes left, Sphere, Box, Box + 2 Pan (box shape with two solar panels), Cyl (cylinder), Cone, and Box + 2 Ant (box shape with two antennas) are the most present shapes in LEO. Together, they represent 84.24% of the total amount of small objects in LEO. Table 1 summarizes the main shapes of small objects found in LEO with their mass $\leq$ 100 kg at the time of writing this paper.

Table 1. Main debris shapes found In LEO (mass $\leq$ 100 kg).

Shape	Amount	% of LEO Small Debris Total
Sphere	1044	25.08
Box	949	22.80
Box + 2 Pan	669	16.07
Cyl	457	10.98
Cone	274	6.58
Box + 2 Ant	113	2.72

On the other hand, Figure 2 shows the distribution of objects in LEO (mass $\leq$ 100 kg) grouped by their shape. Each dot represents a catalogued object relative to its apogee. The shape feature is ordered in descending order, where the sphere shape is the most present, and the Box + 2 Ant satellite shape is the least present.

The data analysis shows that, despite the wide variety of shapes, one generic shape is predominant in LEO: the Box shape (with or without solar panels or antennas). If our capturing mechanism targets all the different Box-shaped objects with mass $\leq$ 100 kg that exists in LEO (Box, Box + 2 Pan, Box + 2 Ant), it will have a clear impact on the debris problem at LEO. Indeed, the Box-shaped objects represent 41.59% of the total amount of small catalogued objects in LEO. Thus, actively catching Box-shaped debris helps answer the problem. Nano-satellites and mega-constellations are the future of LEO exploitation and will quickly saturate LEO. It is then essential to remove those satellites, even before the 25 years of maximum stay in LEO proposed by Inter-Agency Space Debris Committee (IADC) guidelines [42].

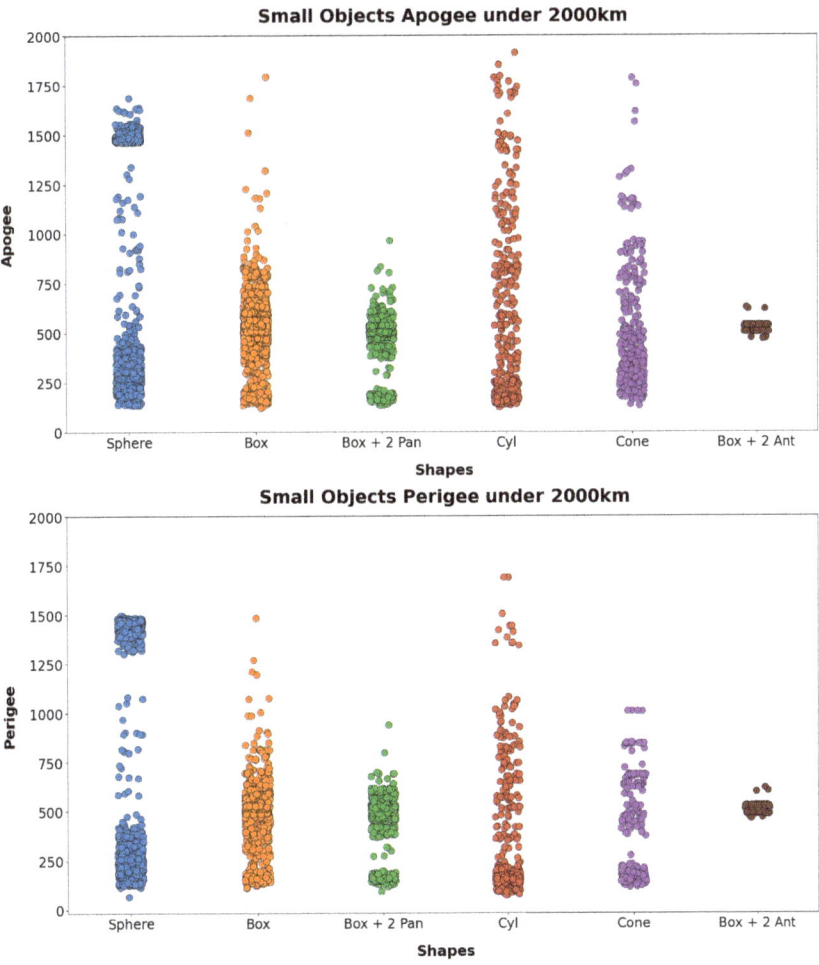

Figure 2. Small objects' apogee and perigee distribution in LEO organised by main shapes.

The hybrid-compliant (the combination of passive and active) system proposed in this paper targets Box-shaped debris of various masses, not exceeding the 100 kg threshold. Other shapes can be considered in further work.

2.2. The Capturing Phase

An ADR mission consists of a succession of several crucial phases. From the launch of the spacecraft from Earth to the moment the chaser satellite, coupled with the debris, burns into the atmosphere, five general phases can be noted: berthed standby (the chaser satellite is on board and attached to the hosting platform), ejection (includes the launch of the rocket until the ejection of the payload), Far-Range Approach (arrive at hold point, close enough to the target), capturing, and post-capture (ready to de-orbit).

The capturing phase is the most crucial one. With little cooperation between the servicer and the target (no communication link, no fiducial markers, nor capture interfaces), capturing uncooperative debris is today one of the biggest challenges. Indeed, mission failure and debris generation can occur more easily during that phase and the consequences can be dramatic.

Figure 3 describes the concept we propose for the capturing phase. It includes three sub-phases, pre-capture (approach guidance and control), soft-capture, and hard-capture; these are in charge of the approach preparation, the impact absorption and stabilization, and the securing of the debris attachment.

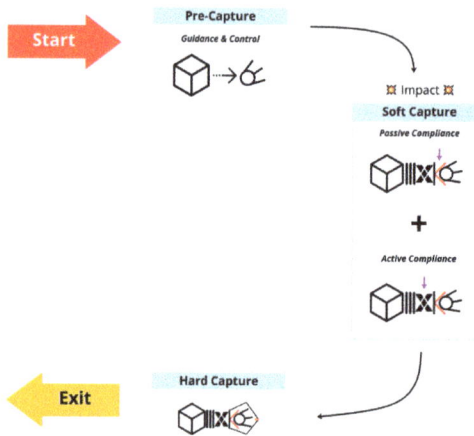

Figure 3. Concept of operations of our proposed capturing phase.

- **Pre-Capture**

The servicer satellite's guidance navigation and control (GNC) rendezvous and synchronizes its motion with the debris. The ADR system is, at first, undeployed inside the CubeSat architecture, as displayed in Figure 4, and is then deployed. At the end of the pre-capture approach, there is a relative distance d_t. Thus, only a translation motion is required to capture the debris.

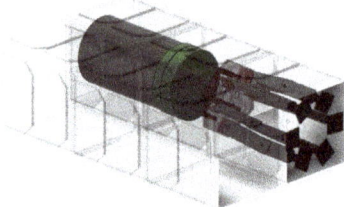

Figure 4. Undeployed hybrid-compliant system. *The gray cylinder at the back of the system is for illustration purposes only.*

- **Soft Capture**

In this sub-phase, the servicer satellite's thrusters are turned on to approach the debris and achieve the first contact. The first impact between the capturing mechanism and the debris must occur softly. Because of this, we propose a hybrid-compliant system for soft capture. It combines passive and active compliance with components that will reduce shocks and residual vibrations and actively control the contact time to avoid motion-reaction effects. It is assumed that the mechanism's tip will remain in contact with the debris for a finite time t_c, long enough to ensure that the hard capture mechanism secures the debris.

- **Hard Capture**

This sub-phase aims to secure the link between the servicer satellite and the debris, resulting in a reliable bond ready for deorbiting. After the soft capture, the hard capture mechanism will activate to fold and embrace the shape of the debris.

Figure 5 presents a general view of the proposed hybrid-compliant system for soft capture integrated into the CubeSat frame. This paper focused on the soft capture sub-phase. Details of the pre-capture and hard-capture sub-phases are out of the scope of this paper. Indeed, the system being at an early-stage design, we assume that the motion synchronization between the servicer and the debris had already been established in the pre-capture phase. Besides, the de-orbiting phase is considered out of the scope of the paper, as it is up to the servicer satellite using the proposed concept to decide how to demise the whole system with the debris attached.

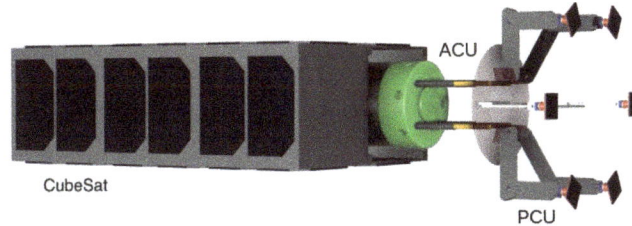

Figure 5. General view of a CubeSat-based hybrid-compliant system for soft capture of space debris. It includes the active compliance unit (ACU) and the passive compliance unit (PCU).

3. Hybrid-Compliant System for a Soft Capture of Space Debris

The high demand for reliability while capturing uncooperative debris makes the system be designed with compliance in mind first. A soft capture at the impact will ensure that the debris is not pushed away and give enough contact time for the capture. To guarantee this soft capture, this paper proposed a hybrid-compliant system at a conceptual level, with passive and active compliance, while fitting into a CubeSat architecture and considering the capturing of uncooperative box-shaped debris in LEO. Figure 6 presents a conceptual close-up view of the hybrid mechanism proposed for the soft capturing sub-phase. It comprises two crucial parts: the active compliance unit (ACU), with tunable stiffness, and the passive compliance unit (PCU), with a permanent stiffness. Together they form the soft capture Uunit (SCU) of adjustable stiffness of our capturing mechanism.

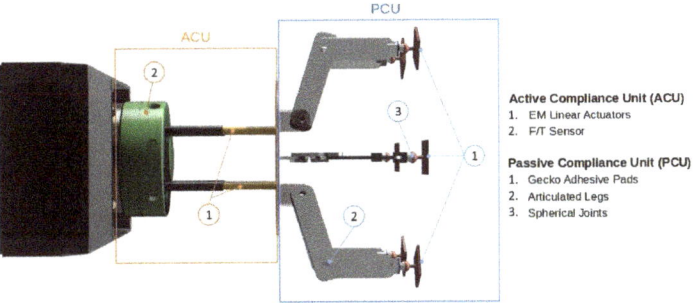

Figure 6. Closeup of the hybrid-compliant system for soft capture of space debris.

3.1. Passive Compliance Unit (PCU)

The PCU, as shown in Figure 6, has two main functions: to ensure a softer impact with the debris, as well as to adhere to the debris surface, both preventing it from moving away. This unit comprises three items: six articulated legs, six spherical joints, and six adhesive pads. The choice of having six legs lies in finding the right balance between the geometry of the system, its weight, and reliability, as fewer legs would question the system's redundancy. This part is the first of the whole system to encounter the target's surface.

- **Articulated Legs**

Each articulated leg, as shown in Figure 7, is composed of three aluminium parts linked together by revolute joints: the lower leg, the upper leg, and the link between the leg and the plate. The latter separates the PCU and the ACU. The passive compliance and flexibility feature is then made possible thanks to torsional springs located in the joints of the legs. The choice of adding torsional springs is for two reasons; to have a softer impact and better safety concerns regarding the system's integrity by avoiding high compression and bending constraints.

The torsional springs' stiffness will determine the PCU's maximum displacement in the axis of capture. This parameter plays an important role in the design of the overall hybrid compliance of the soft capture. Depending on the debris parameters (such as its mass), a too-low stiffness of the PCU could result in a longer displacement of the legs, and as a result, the system could break under the generated constraints. This point will be discussed later in Section 5.

- **Spherical Joints**

To link the legs to the adhesive pads, spherical joints are integrated. They give the pads two more mechanical degrees of freedom (in our case, two free rotations and the rotation in the axis of capture blocked), thus the better possibility to adapt to the debris surface. Indeed, in the case of a slight misalignment between the chaser satellite and the debris surface, the adhesion might not occur. In that regard, ensuring the parallelism of the pads with the debris surface is of utmost importance for efficient adhesion [43].

- **Gecko Adhesive Pads**

At the tip of each leg, a gecko-inspired dry adhesive [44] component is integrated as a thin layer under the pads. This dry, yet sticky, material must be activated by applying a shear force [43]. As a result, the microscopic "hairs" bend, creating a wider contact area between the pad and the target's surface, then making adhesion possible to many different material surfaces. These pads would also include a contact sensor [31] so that the control algorithms know exactly when to activate the adhesives. The shear force is created thanks to the active shrinkage of the legs towards the capture axis. This bio-inspired dry adhesive fits well for our case for two main reasons. Firstly, adhering within the required contact time to the debris surface is one way to avoid the action-reaction effect while creating sufficient time for securing the debris-chaser link. Moreover, selecting a dry adhesive that requires shear force to activate fulfils some of the requirements for this concept: besides being able to be used in a space environment [30], no additional normal force is required to adhere. Indeed, applying more contact force when one tries to avoid pushing the debris away sounds paradoxical. To that extent, getting a dry adhesive activated by shear force is the most suitable solution for catching space debris more reliably.

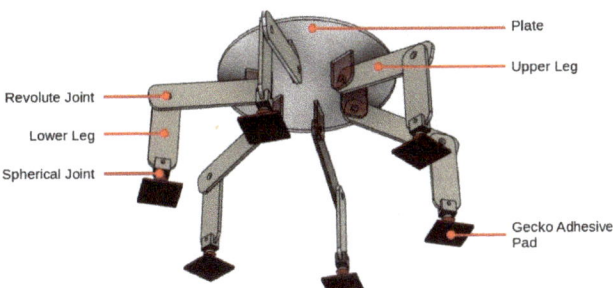

Figure 7. General view of the PCU, composed of one plate, six articulated legs, and six gecko-inspired dry adhesive pads.

3.2. Active Compliance Unit (ACU)

The ACU, as shown in details in Figure 8, is directly linked to the PCU with the same plate shown in Figure 7. This unit comprises four linear actuators linked with their base to a force/torque (F/T) sensor. Active compliance is ensured thanks to the active control of the linear actuators along the capture axis. Details about the controller are presented in Section 4. By actively changing the stiffness of the ACU, it is possible to ensure a sufficient contact time to actuate the other parts of the ADR capturing process. This allows the system to target a wider range of debris without fundamentally changing its conceptual design.

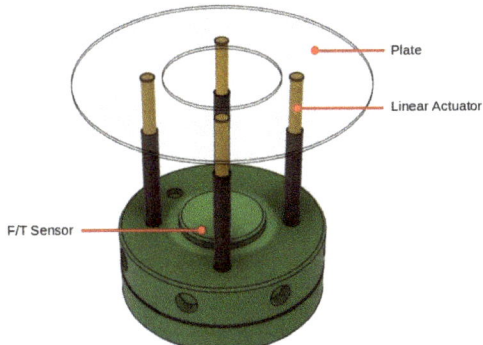

Figure 8. General view of the ACU. The CDU is composed of one force/torque sensor, four electromechanical linear actuators and one plate.

- **Force/Torque Sensor**

The presence of an F/T sensor, as seen in green in Figure 8, helps feed the controller with the force encountered at the impact between the chaser satellite and the debris. Consequently, the linear actuators will be actuated regarding the force sensed by the F/T sensor, providing the required equivalent stiffness of the overall system towards the targeted debris. This means one can change the parameters of the stiffness and damping of the ACU.

- **Electromechanical Linear Actuators**

The four linear actuators are an essential part of the ACU. They are represented in Figure 8 with the static part in gray and the dynamic part in yellow. Although a single linear actuator would have performed the task properly, a failure in that system can generate mission failure. In that regard, it is important to ensure better reliability of that part of the system by using four redundant electromechanical linear actuators. The linear motion of the actuators makes them act as a spring and damper system in a controlled way. As a result, active compliance is created with the F/T sensor in a control loop.

3.3. Soft Capture Process

Both the ACU and the PCU will work together towards a successful soft capture of the space debris, as described in Section 2.2. As a reminder, the main goal of the soft capture sub-phase is to absorb the impact and welcome, as softly as possible, the debris while retaining it from moving away. During the capture process, the actions of the soft capture can be depicted in four main steps, as displayed in Figure 9: the Initialization, the first contact, the hybrid compliance operation (active and passive compliance occurs simultaneously), and the adhesive activation. Video S1 attached to this paper provides a visual understanding of the described steps.

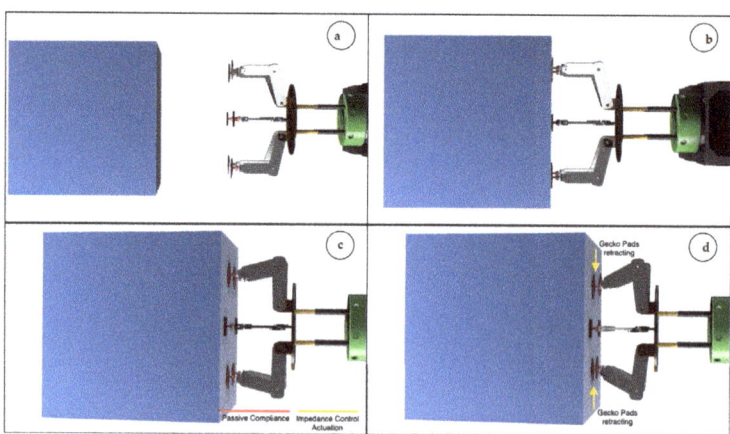

Figure 9. Soft capture process of the proposed ADR concept: (**a**) initialisation; (**b**) first contact; (**c**) hybrid compliance operation; (**d**) adhesive activation.

- **Initialisation**

At this moment of the process, the ADR system is already deployed, and only a relative distance d_t separates the servicer satellite from the debris. The servicer satellite approaches the debris with a translation motion, as depicted in the first image in Figure 9a.

- **First contact**

The PCU is the part which arrives in contact with the flat surface of the debris first, with its gecko adhesive pads parallel to the debris surface, as shown in Figure 9b.

- **Hybrid Compliance Operation**

As the contact is made, the flexible legs articulate instantly, as seen in Figure 9c, providing the first damping of the impact's vibrations and not being too close to an elastic collision between the two entities (where both momentum and kinetic energy are conserved). The fixed stiffness of the PCU lets the legs articulate while keeping in contact with the debris surface.

At this time of the process, the PCU is not the only one acting; the ACU is also activated at the impact. As soon as there is contact between the SCU and the debris, the force exerted in the axis of capture on the ADR system's tip is fed into the controller of the ACU. As a result, the electromechanical linear actuators are put into action accordingly, reducing their length (as shown in Figure 9c) and thus providing an additional set of virtual springs and dampers based on the contact's force. The contact time t_c between the tip of the capturing system and the debris can then be controlled thanks to the ACU.

- **Adhesive Activation**

    The action of passive and active compliance is performed within that time frame of t_c seconds, giving the required theoretical time for the adhesive activation to occur, which is essential to the mission's success. The last goal of the soft capture sub-phase (retaining the debris from moving away due to the action-reaction effect) is made possible by creating adhesion on the debris surface. Within the contact time t_c, the adhesive pad's contact sensors must send a positive signal to the ADR system's process controller and activate the pads' shrinkage, as shown in Figure 9d. Working in opposite pairs, the pads are pulled towards the longitudinal axis of the capture, towards the centre of the ADR tip's plane; shear force is necessary to adhere. That shear force is maintained to keep the adhesives activated, retaining the debris from moving away.

    Once the bond is created between the servicer and the debris, the soft capture sub-phase is performed, and the capture phase can proceed.

4. Impedance Controller

To actively remove space debris, a servicer CubeSat will have to perform the final approach, deploy the dedicated mechanism, and then perform the capturing phase of the ADR mission. Having a hybrid-compliant system implies that both *passive* and *active* compliance are involved. Since the passive compliance has fixed stiffness and damping coefficients, it is required to analyse and model the adequate controller to get the active compliance's right coefficients. The CubeSat and the debris are specific in mass, but the capturing mechanism's compliance can be modified for the optimal response of the ADR system regarding the contact time with the debris. In this section, the aim was to study the behaviour of the systems during contact and then regulate the relationship between the ADR system's tip and contact force, employing an impedance controller. A single-axis analysis was undertaken (central impact), as is common in the literature [45].

4.1. System Modeling

The servicer satellite, consisting of a main body (CubeSat), and the hybrid compliant system for soft capture, is modelled as a three-body equivalent system, as represented in Figure 10. The CubeSat, along with the ACU's F/T sensor and the fixed part of the ACU's linear actuators, are lumped into the first rigid body with mass m_s. The moving part of the ACU's actuators, the plate that separates ACU and PCU, and the PCU's upper legs are lumped into a second rigid body with mass m_e; while the lower legs and the gecko adhesive pads are lumped to a third rigid body having mass m_c. The positions of the center of mass (CoM) of m_s and m_e are denoted by x_s and x_e, and the position of the mechanism's tip is denoted by x_c. The debris is modelled as a rigid body of mass m_d, and the position of the point on the debris that comes into contact with the mechanism's tip is denoted by x_d.

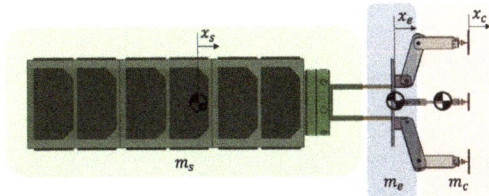

Figure 10. Equivalent three-body system of the CubeSat and ADR system.

Masses m_s and m_e are connected through linear actuators, allowing a translation degree of freedom to be controlled. The maximum displacement of the linear actuators' moving parts is denoted by l_a. Masses m_e and m_c are connected through passive compliance, with stiffness k_s and damping b_s, that models the compliance provided by the 6 torsional springs located in the revolute joints of PCU's legs shown in Figure 7. Figure 11 provides a simplified 2D view of the three-body system.

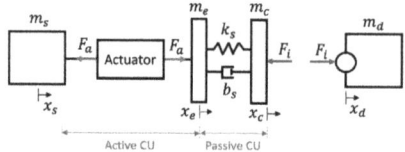

Figure 11. Schematic of the equivalent three-body CubeSat-ADR system.

Before the contact, the CubeSat-ADR system has a non-zero relative velocity with respect to the debris. Once the PCU's mass m_c arrives in contact with the flat surface of the debris mass m_d at the moment t_i, m_c and m_d have the same position, i.e., $x_c = x_d$, the passive compliance enters into motion instantly, and the impedance controller is activated.

The aim of the simulation study was to showcase the importance of incorporating active and passive compliant components to dissipate impact energy, ensure contact time, and enhance the capture of a wider range of space debris masses. Therefore, the simulation study was based on the following assumptions. The motion synchronization between the servicer and the debris was established, resulting in a zero relative angular velocity. The desired contact point of the ADR system's tip on the debris was assumed to pass through the debris centre of mass, resulting in only a contact force and no external moment on the debris. The assumption was made that the centre of mass of the debris is known, supported by existing research on estimation techniques. Misalignments during realistic approach and contact were not considered and flat surfaces were assumed for both the ADR system and debris, generating contact force along the approach and contact axis.

Based on these assumptions, a three-dimensional simulation of the equivalent three-body system yields single-axis motion for the servicer and the ADR system was performed, providing informative data along the approach and contact axis. Due to the absence of relative rotational motion, all motion occurs along this axis. The inclusion of the assumption of point masses in the simulation model, neglecting the moment of inertia, does not affect the study's conclusions. The paper presents the equations of motion for this equivalent system, focusing on the commonly employed central impact analysis of the single motion axis.

Specifically, the system equations of motion for each of the three rigid bodies of the equivalent CubeSat-ADR system in Figure 11 with masses m_s, m_e, and m_c, and for the space debris with mass m_d, obtained during the contact between m_c and m_d, are given by Equations (1), (2), (3) and (4), respectively.

$$m_s \ddot{x}_s = -F_a \tag{1}$$

$$m_e \ddot{x}_e = F_a + k_s(x_c - x_e - l_s) + b_s(v_c - v_e) \tag{2}$$

$$m_c \ddot{x}_c = -F_i - k_s(x_c - x_e - l_s) - b_s(v_c - v_e) \tag{3}$$

$$m_d \ddot{x}_d = F_i, \tag{4}$$

where F_a is the commanded force applied on the capture unit by the impedance-controlled linear actuator, l_s is the physical length of spring k_s, and F_i is the impact force between the mechanism's tip and the debris. All forces are shown in Figure 11.

4.2. Design of the Impedance Controller

For successful adhesion, the required contact time between the ADR system's tip and the debris must be ensured; thus, its adjustment is required. This adjustment was achieved by altering the ADR system's impedance. Therefore, an impedance controller with tunable gains was developed. Specifically, impedance control attempts to implement a dynamic relation between the ADR system's variables, such as tip position and contact force, rather than just controlling these variables alone [37].

Subsequently, the controller needs to be informed, which is the wanted relation between the ADR system's variables during impact i.e., the desired system's behavior. The equation selected to describe this behavior is called *impedance filter* and is shown in Equation (5), [34,46]. It consists of three terms: one for the desired inertia m_f to be seen at the tip, one for the desired damping b_f, i.e., the desired relationship between contact force and tip's velocity, and one for the desired stiffness k_f, i.e., the desired relationship between contact force and tip's displacement [37].

$$m_f(\ddot{x}_c - \ddot{x}_s) + b_f(v_c - v_s) + k_f(x_c - x_s - l_m) = -F_i. \quad (5)$$

The desired contact time of the ADR system with the debris and, thus, the success of capturing directly, can be realized by tuning the mass, spring, and damper impedance parameters m_f, b_f, and k_f, respectively. Parameter l_m in Equation (5) is the initial distance between m_c and m_s.

Substituting $\ddot{x}_s$ of Equation (1) and $\ddot{x}_c$ of Equation (3) into the impedance filter in Equation (5), and then, solving for the applied actuator force by the impedance controller F_a required to achieve the desired impedance behavior of Equation (5), yields

$$F_a = \frac{m_s}{m_f}(\frac{m_f}{m_c} - 1)F_i + \frac{m_s}{m_f}k_f(x_s - x_c + l_m) + \frac{m_s}{m_f}b_f(v_s - v_c) + \frac{m_s}{m_c}k_s(x_c - x_e - l_s) + \frac{m_s}{m_f}b_s(v_c - v_e). \quad (6)$$

The impedance parameter m_f is selected equal to m_c so that the actuator force F_a does not depend on the impact force F_i [34]. Then, the applied actuator force F_a is given by

$$F_a = k_p(x_s - x_c + l_m) + k_d(v_s - v_c) + \frac{m_s}{m_c}k_s(x_c - x_e - l_s) + \frac{m_s}{m_c}b_s(v_c - v_e), \quad (7)$$

where the controller's gains k_d and k_p are given by

$$k_p = \frac{m_s}{m_c}k_f \quad (8)$$

$$k_d = \frac{m_s}{m_c}b_f. \quad (9)$$

To calculate the gains based on Equations (8) and (9), the impedance parameter k_f must be selected. Furthermore, choosing critical damping results in the impedance parameter b_f.

$$b_f = 2\sqrt{m_f k_f}. \quad (10)$$

The impedance control loop is shown as a block diagram in Figure 12.

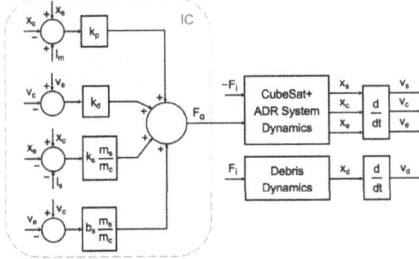

Figure 12. Block diagram of the impedance control strategy for the soft capture of space debris.

4.3. Hybrid Compliance

The useful terms of *active* and *hybrid* compliance are described in this section to understand the proposed impedance controller better. For this purpose, we used a reduced version of the three-mass system previously described in Figure 11. The reduced version is a two-mass equivalent ADR system with only the ACU to control its interaction with the debris, as presented in Figure 13a.

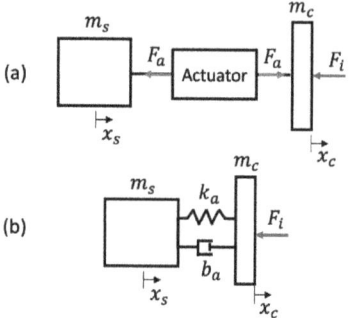

Figure 13. CubeSat-ADR system with two masses connected by (a) impedance-controlled actuators, (b) an active (virtual) compliance equivalent to (a).

The system equations of motion for the equivalent ADR system in Figure 13a, during the contact with the debris, can be written as:

$$m_s \ddot{x}_s = -F_a \tag{11}$$

$$m_c \ddot{x}_c = -F_i + F_a. \tag{12}$$

Substituting $\ddot{x}_s$ of Equation (11), and $\ddot{x}_c$ of Equation (12), into the impedance filter that describes the desired impact behavior of Equation (5) yields

$$-\frac{m_f}{m_c}F_i + \frac{m_f}{\mu_{ef}}F_a + b_f(\dot{x}_c - \dot{x}_s) + k_f(x_c - x_s - k_m) = -F_i, \tag{13}$$

where μ_{ef} is given by

$$\mu_{ef} = \frac{m_c m_s}{m_c + m_s}. \tag{14}$$

Solving for the actuator force F_a and selecting m_f equal to m_c so that F_a does not depend on the impact force F_i, yields

$$F_a = k_p(x_s - x_c + l_m) + k_d(\dot{x}_s - \dot{x}_c), \tag{15}$$

where k_d and k_p are the impedance controller's gains given by

$$k_d = \frac{\mu_{ef}}{m_f}b_f \tag{16}$$

and

$$k_p = \frac{\mu_{ef}}{m_f}k_f. \tag{17}$$

Observing Equation (15) for the actuator's force command F_a, one can conclude that the impedance-controlled actuator behaves in active (virtual) compliance with spring coefficient k_a of length l_a and damping coefficient b_a, as shown in Figure 13b; in this example, equal to the controller's gains k_p and k_d, respectively.

Furthermore, the proposed ADR system, as modelled in Section 4.1 and shown in Figure 11, incorporates, additionally to the active compliance, passive physical compliance with spring coefficient k_s of length l_s and damping coefficient b_s, see Figure 14a.

The active and the passive compliance in series can be combined to form an equivalent *hybrid* compliance of length $l_m = l_s + l_a$ with spring coefficient k_m and damping coefficient b_m as shown in Figure 14b.

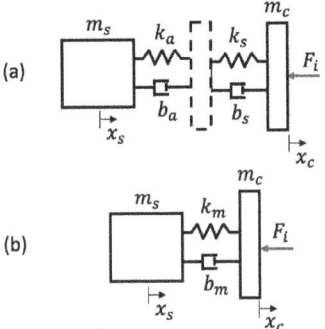

Figure 14. CubeSat-ADR system as two masses connected by (**a**) a passive and active compliance in series, (**b**) a hybrid compliance, equivalent to (**a**).

The stiffness and the damping coefficients k_m and b_m of the hybrid system are of paramount importance as they affect the ADR system's impedance, the contact time of the ADR system with the debris and, thus, the success of the debris capture. Therefore, the reduced hybrid-compliant system shown in Figure 14b is used in Section 5.4 to showcase the necessity of hybrid compliance in an ADR system.

5. Simulation Study and Results

A series of simulations were conducted with three objectives in mind: to study the relationship between the debris mass and the required compliance and to demonstrate the importance of the proposed hybrid compliant system (Section 5.2); to study the impact of the design parameter l_m (Section 5.3); and to test the impedance controller and analyze its role to achieve a soft capture of space debris (Section 5.4).

5.1. Simulation Setup

The simulations were run in MATLAB/Simscape using a variable-step ode45 solver. The Simscape model, consisting of the hybrid-compliant system mounted on the servicer CubeSat and the space debris, were developed for the simulations. During the simulations, the positions and velocities of the masses under the impact and their interpenetration were calculated. This was fed back to a contact model and a force was generated, pushing away the masses under impact. The contact time t_c was calculated based on the impact force.

The developed contact model uses the visco-elastic theory. According to this theory, a compliant surface under impact can be modelled by a combination of lumped parameter elements, i.e., springs and dampers. This study calculated the contact force between the bodies under impact using the Kelvin–Voight model [47]. Assuming that the impact is close to an elastic (no damping), the impact force is given by:

$$F_i = k_i(x_c - x_d), \tag{18}$$

where x_c is the position of the mechanism's tip and x_d is the point on the debris that comes into contact with the mechanism's tip. In this study, stiffness k_i was equal to 10,000 N/m [48,49] and, thus, the contact was assimilated to a very stiff spring, activated when x_c is greater than x_d.

The CubeSat-ADR system has a small relative velocity set to 10 mm/s with respect to the debris. The CoM's initial position x_s, of mass m_s, equals zero before impact. The initial position of x_c equals l_m (l_m is defined in each experiment). The debris' initial position relative to the ADR system's tip, denoted by $x_d - x_c$, was set equal to 10 cm without loss of generality since, in the simulation, the ADR system approaches the debris with a constant velocity v_s. Equivalent systems' point masses m_s, m_c, and m_e (when applicable) are 12.0012 kg, 0.024 kg, and 0.016 kg, respectively.

5.2. Debris-Mass and Compliance Relation

As the masses of the servicer CubeSat, including the ADR system, were assumed to be known, the desired stiffness and the damping coefficients of the hybrid system's equivalent compliance must be selected. The selected parameters should ensure that the minimum contact time between the ADR system and the debris was achieved. An analytical solution for the optimal tuning of these coefficients is difficult to obtain since no analytical equation relates the contact time and the hybrid compliance coefficients. Because of this, an algorithm in MATLAB, consisting of a loop, was developed to search the successful cases (t_c > minimum contact time required) in a range of stiffness values, for a range of space debris masses and for a range of minimum contact time required to complete the capture.

Specifically, the servicer CubeSat and the ADR system were simulated when approaching and coming into contact and the success in terms of the time of contact was noted. It was considered a successful case if it was greater than the contact time required for the successful capturing while not reaching the spring limit. Then, the corresponding spring's stiffness and the debris mass were stored.

In this simulation study, for tuning the hybrid compliance of the system, the servicer CubeSat and the ADR system were modelled as a two-mass equivalent system, i.e., as two point masses connected by the hybrid compliance, as shown in Figure 14b. This compliance was considered hybrid since it consists of passive parts integrated into the PCU and the active part realized by the impedance-controlled linear actuator of the ACU, as shown in Figure 14a.

The stiffness and damping coefficients to be altered during the search of the developed algorithm are denoted by k_m and b_m, respectively. Once k_m is altered, by choosing critical damping, one can calculate b_m too, as follows

$$b_m = 2\sqrt{m_c k_m}. \tag{19}$$

The hybrid compliance's length l_m is equal to 0.05m since it is the sum of two lengths: $l_m = l_s + l_a$; the length l_s of the passive physical compliance, equal to 0.025 m, and the maximum displacement l_a of the linear actuators' moving parts, equal to 0.025 m. The schematic of the system under simulation study as designed in Simscape is shown in Figure 15.

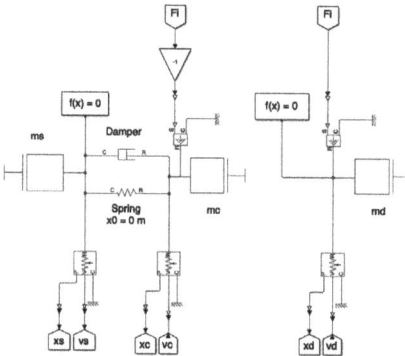

Figure 15. Schematic of CubeSat-ADR system and debris in Simscape.

The algorithm runs for the range of stiffness values k_m between [0.1–0.5] N/m, with a step of 0.1 N/m, for a range of space debris mass m_d between [1–100] kg, with a step of 1 kg, and for a range of minimum contact time t_c required for completion of the capturing between [4–12] s, with steps of 2 s.

The resulting diagram is shown in Figure 16, providing the relation between these variables. More detailed visualization of the data of Figure 16 is provided for each contact time in the range of [4–12] s in Appendix A. The lines depicted in the 3D diagram correspond to points in 3D space, representing the successful cases obtained from the algorithm. Each point along the line represents three distinct values, namely: space debris mass, minimum contact time achieved, and hybrid compliance stiffness coefficient.

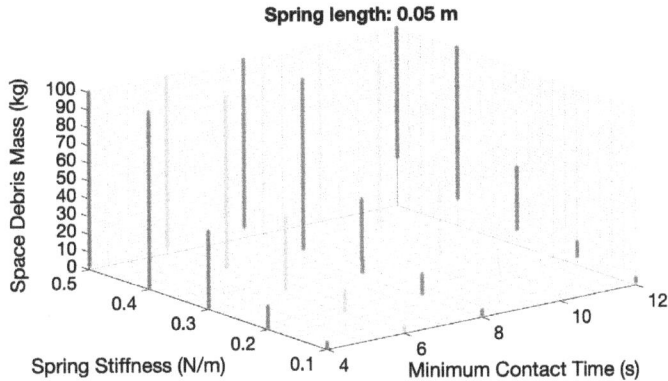

Figure 16. Range of space debris masses to be targeted for different stiffness coefficients and minimum contact times achieved.

Note that the [4–12] s range for this simulation for the minimum contact time required was selected since it was sufficient to showcase the necessity of tuning the system's compliance and, therefore, of the hybrid compliance concept. Simulation results for minimum contact time required greater than 12 s show that hybrid compliance is even more necessary if we want to target a wide range of debris between 0–100 kg. This can be easily shown by the trend shown in Figure 16: the minimum contact time required increases, and the range of debris masses to be targeted decreases. Moreover, a contact time of less than 4 s for successful capturing is considered unrealistically small.

Based on the diagrams, desired stiffness and damping coefficients of the hybrid system's equivalent compliance can be selected for a specific debris mass and minimum contact time required.

In Figure 16, the relation of the variables is derived. In particular, when the equivalent stiffness increases for a specific minimum contact time, the range of the debris masses increases and the minimum debris mass to be targeted increases. One could say that small stiffnesses are appropriate for targeting a small range of small debris and larger stiffnesses are appropriate for targeting a wider range of debris masses of larger debris masses. Furthermore, when the minimum contact time required increases, the maximum debris mass to be targeted remains constant for a specific spring stiffness due to displacement limitations of the compliant parts and the minimum debris mass that can be targeted increases. Thus, when the minimum contact time required increases, the range of debris masses to be targeted decreases.

5.2.1. Indicative Case

In this section, indicative plots are presented based on the simulation responses for a specific set of values in the range searched by the algorithm. Specifically, the indicative plots in this section were obtained using a simulation with k_m and b_m equal to 0.5 N/m and 0.1789 Ns/m, respectively, and debris with mass m_d equal to 12 kg. The positions of the point masses m_s, m_c and m_d are shown in Figure 17. As shown in this figure, the tip of the ADR system was initially located 10 cm from the debris. For almost 10 s, the ADR system approaches the debris, moving together while in contact.

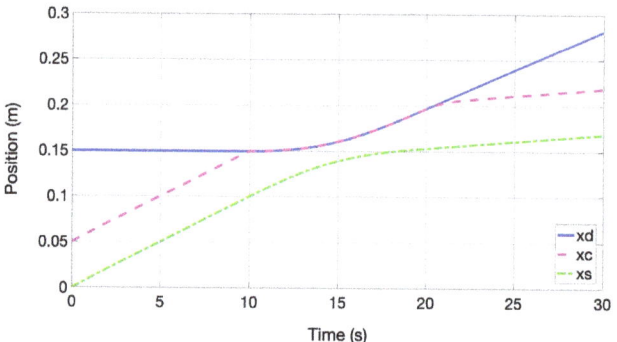

Figure 17. Position of masses for the indicative case.

The contact time t_i was calculated using the impact force shown in Figure 18. It is the time when the impact force is continuously greater than zero and, thus, is equal to 10.12 s. The impact force in Figure 18 was set to zero when the relative position of the ADR system's tip from the debris, shown in Figure 19, was negative, indicating that there is a distance between the two systems. However, when the systems are in contact, the interpenetration of the bodies, shown in Figure 19, is positive, and the impact force was calculated based on Equation (18); it is the multiplication of the spring stiffness k_i times the interpenetration in Figure 19.

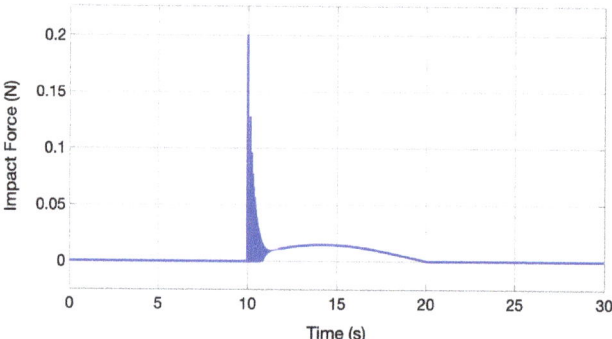

Figure 18. Impact force for the indicative case.

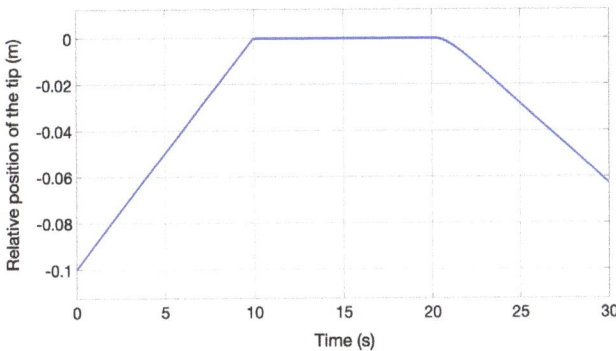

Figure 19. Relative position of ADR system's tip with regard to the debris contact point for the indicative case.

5.2.2. Discussion: The Need for Hybrid Compliance

The analysis of the diagram in Figure 16 leads to a conclusion of major importance regarding the design of the ADR system itself. When the required minimum contact time is very small, there is indeed an appropriate stiffness coefficient for targeting most debris masses in the desired range of 0–100 kg. However, for more realistic minimum contact times required, none of the stiffness coefficients are adequate; one must be able to modify the equivalent spring's stiffness to target the whole desired range of debris.

Considering the use case where the system only uses the benefits of passive compliance, the equivalent stiffness would remain constant without any possibility of being tuned. In that regard, the range of debris that can be targeted is consequently constrained. Assuming a passive spring of 0.5 N/m, and the required minimum contact time is 10 s, based on the diagram in Figure 16 and the more detailed visualization of it provided in Figure 20, the range of debris masses that can be targeted is 12–100 kg. Nevertheless, to capture debris of smaller mass, e.g., 4 kg, an equivalent stiffness coefficient of 0.2 N/m would be required for the equivalent spring, as shown in Figure 20. To achieve the tuning of the spring's stiffness from 0.5 N/m to an equivalent spring's stiffness of 0.2 N/m, active compliance should be added, realized by an impedance controller, adding versatility to the system.

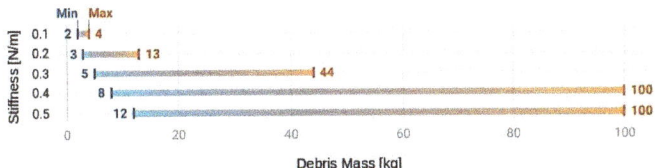

Figure 20. Detailed visualization of Figure 16 for minimum contact time achieved of 10 s. Range of debris masses for different stiffness coefficients.

To decrease the equivalent spring stiffness k_m to make it equal to 0.2 N/m, the additional active compliance k_a should be in series with the already-manufactured and integrated passive one k_s, with k_s equal to 0.5 N/m in this study. Hence, based on Equation (20) for springs in series, the stiffness coefficient of the active compliance k_a should be equal to 0.333 N/m. Employing this active compliance in the presence of the passive one, the debris of 4 kg can be successfully captured, thus allowing the ADR system to target a wider range of debris than the one targeted by employing only the passive compliance.

$$k_m = \frac{k_s k_a}{k_s + k_a}. \tag{20}$$

One could wonder why not use only active compliance. For reliability reasons, integrating flexibility with passive compliance at the first impact interface would avoid a hard shock and, thus, avoid pushing away the debris. Moreover, if the debris has an unexpected mass variation, tuning the stiffness brings more reliability and safety, reducing the risk of damaging either the servicer or the debris itself.

5.3. Impact of the Compliance's Physical Length

The series of simulations and results are presented in Figure 16 and in Appendix A, considering a length l_m of the equivalent spring equal to 0.05 m. The physical length of the spring denotes the available space of the mechanism to compress, as shown for the indicative case in Figure 21. It is, therefore, an important design parameter for the system.

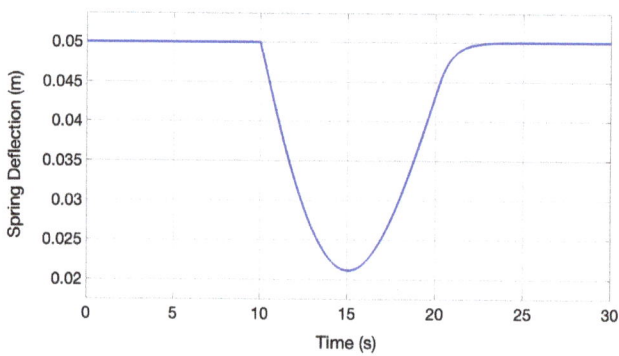

Figure 21. Spring deflection for the indicative case.

A series of simulations were run for a different spring's length to further enhance the discussion and the valuable conclusions. The algorithm developed searches for the range of stiffness k_m between 0.25–1.25 N/m, for a range of space debris mass m_d between 1–100 kg and for a range of minimum contact time required for completion of the capturing between 5–8 s. Specifically, Figure 22 displays the range of space debris masses that can be targeted and successfully captured for a range of equivalent spring stiffnesses and for various minimum contact times required for a spring's length l_m equal to 0.025 m.

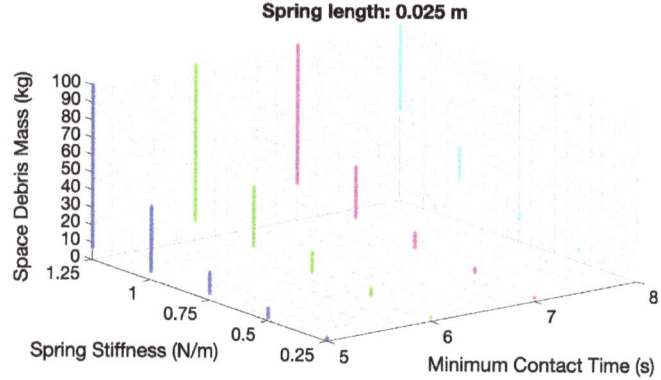

Figure 22. Range of space debris to be targeted for different contact times.

Based on this diagram, one could observe that, for a specific minimum contact time and spring's stiffness, the range of debris masses that can be targeted is smaller than the corresponding one when the spring's length l_m is equal to 0.05 m. This is because, at similar stiffnesses and contact time parameters, the retracting phase on the spring reaches its limit, resulting in possible damage to the servicer CubeSat. In other words, the contact time requirement may be fulfilled while reaching the spring's length limits, which may be dangerous for the servicer satellite. Hence, the ability to tune the equivalent stiffness coefficient using an impedance controller is even more necessary. Moreover, as shown in Figure 22, the maximum contact time achieved is no more than 8 s; this is an important contact time constraint. In conclusion, for an ADR system to target a wider range of debris masses while ensuring a realistic required time of contact, the spring's length for compliance—or alternatively the space that the system has to compress—must be carefully selected to be above a minimum value.

5.4. Evaluation of Hybrid Compliance

Two versions of the ADR system were simulated to further demonstrate the importance of tuning the ADR system's compliance by adding an active compliant unit and to showcase the application of the proposed impedance controller. Subsequently, the responses were measured and the corresponding contact times were calculated and compared.

The first system is a passive system, denoted as PCS, composed only of passive physical compliance. In this case, the ACU is inactive; thus, its prismatic joints are locked, see Figure 23a. The second system is a hybrid-compliant system, denoted as HCS and shown in Figure 23b, composed of passive and active compliance; the linear actuators apply forces F_a driven by the proposed impedance controller presented in Section 5.2 and given by Equation (7).

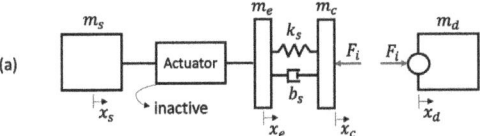

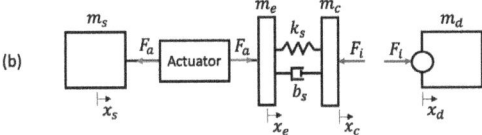

Figure 23. Comparison of ADR systems; (**a**) PCS: system with only passive compliance (inactive linear actuators); (**b**) HCS: hybrid system with both passive and active compliance.

The passive compliance's stiffness k_s equals 0.5 N/m following the use case presented in Section 5.2 and its physical length l_s is equal to 0.025 m. The linear actuator's space limit, i.e., the maximum displacement $l_a = x_e - x_s$ allowed, is also equal to 0.025 m.

Figure 24 shows the impact force generated at the tips of the PCS system (magenta line) and the HCS system (blue line). Figure 25 shows the actuator force commanded by the impedance controller to be less than 0.15 N.

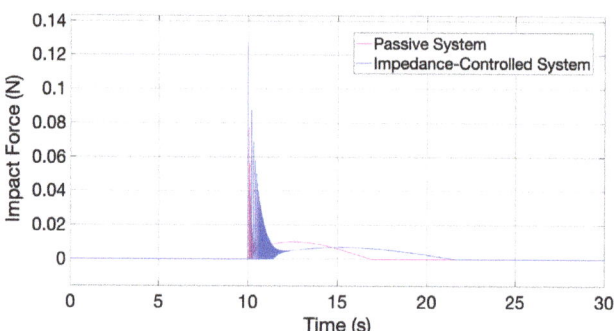

Figure 24. Comparison of impact forces of passive and impedance controlled systems.

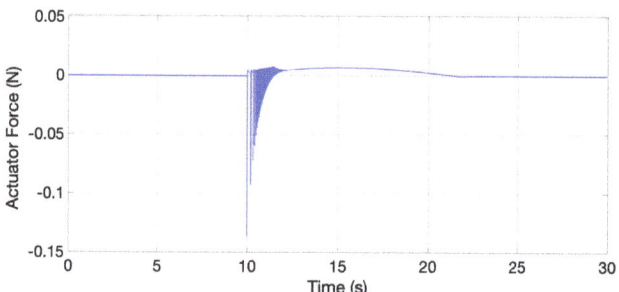

Figure 25. Commanded force by the impedance-controlled linear actuator.

Observing Figure 24 and comparing the contact times, one can see that using the proposed impedance controller significantly increases the contact time from 7 s to 11.65 s. Hence, the applied controller adjusted the contact time between the ADR system's tip and the debris and ensured the minimum required; in this example, equal to 10 s. The desired contact time of the ADR system was ensured by the appropriate tuning of the mass, spring, and damper impedance parameters m_f, k_f, and b_f of Equation (5), respectively, and therefore of the IC gains.

To find the appropriate impedance parameters m_f, k_f, and b_f, the simulation results provided in Section 5.2 for the use case were employed. Specifically, to capture debris with a mass of 4 kg, the required hybrid compliance's stiffness and damping coefficients, k_m and b_m, were found to be equal to 0.2 N/m and 0.17 Ns/m, respectively. Thus, the desired spring and damper parameters of the impedance filter k_f and b_f can be calculated based on Equation (16) as

$$k_f = \frac{m_f}{\mu_{ef}} k_m \qquad (21)$$

and,

$$b_f = \frac{m_f}{\mu_{ef}} b_m, \qquad (22)$$

to be equal to 0.2003 N/m and 0.1702 Ns/m, respectively. The desired mass parameter m_f is equal to m_c. Using the desired impedance parameters m_f, b_f, and k_f derived, the IC gains k_p and k_d were calculated based on Equations (8) and (9) to equal 150 N/m and 128 Ns/m, respectively. The IC command F_a, which drove the ACU's linear actuators, was calculated by Equation (7).

The application of the IC implemented active (virtual) compliance into the ADR system, rendering it a hybrid-compliant system. Comparing the PCS and HCS, the necessity of the IC and, subsequently, of a hybrid-compliant ADR system for the successful capturing of debris, was validated.

6. Conclusions

This paper proposed a one-to-many solution: a flexible, versatile capturing mechanism of class ET2 targeting a vast range of small uncooperative space debris in low Earth orbit (LEO). It incorporates a hybrid-compliant system, combining active compliance (with controlled linear actuators) and passive compliance (with legs articulated by torsional springs). Combined, they make the equivalent hybrid stiffness adjustable to a specific range of debris mass. This novel system also uses a bio-inspired dry adhesive to stick to the debris surface and keep it from being pushed away, increasing the overall reliability of the ADR mission.

The simulation study presented in this paper revealed that a passive-compliant ADR system was incapable of targeting all the small debris. The integration of both active and passive compliance was required to enable the successful soft capturing of the whole range of small debris (up to 100 kg). It allows the system to gently welcome the debris in contact with the servicer satellite, providing the required contact time for properly capturing it. The active compliance is controlled by the developed impedance controller (IC), which adjusts the compliance parameters based on the debris that will be captured.

This paper brings forward the research on capturing a wide range of small debris in orbit, thus contributing to a cleaner and safer space. Future work will focus on the design, development, assembly, verification, and validation (V&V) of all components of the mechanism and experimental V&V testing in the Zero-G Lab facility of SnT-University of Luxembourg.

Supplementary Materials: The following supporting information can be downloaded at: https://www.mdpi.com/article/10.3390/app13137968/s1, Video S1: Simulation video of the soft capture process. Software used: NVIDIA Omniverse.

Author Contributions: Conceptualization, M.H.D. and O.-O.C.-L.; Software, M.H.D. and O.-O.C.-L.; Validation, C.M.; Investigation, M.H.D. and O.-O.C.-L.; Data curation, M.H.D.; Writing—original draft, M.H.D. and O.-O.C.-L.; Writing—review & editing, M.H.D., O.-O.C.-L., B.C.Y., X.L. and C.M.; Supervision, M.O.-M. and C.M. All authors have read and agreed to the published version of the manuscript.

Funding: SnT-SpaceR has conducted this study under Luxembourg National Research Fund (FNR)—BRIDGES funding for "High fidELity tEsting eNvironment for Active Space Debris Removal—HELEN", project ref: BRIDGES2021/MS/15836393, and FNR funding for "design of a Capturing, Absorbing, SEcuring system for active space Debris removal—CASED" project FNR16678722.

Institutional Review Board Statement: Not applicable.

Informed Consent Statement: Not applicable

Data Availability Statement: Restrictions apply to the availability of these data. Data was obtained from DISCOS (Database and Information System Characterising Objects in Space) and are available at https://discosweb.esoc.esa.int/objects with the permission of ESA.

Conflicts of Interest: The authors declare no conflict of interest.

Appendix A

More detailed visualisation of the data of Figure 16 is provided for each contact time in the range of [4–12] s in Figures A1–A4 and 20. Based on the derived diagrams, the desired stiffness and damping coefficients of the hybrid system's equivalent compliance can be selected for a specific debris mass and minimum contact time required.

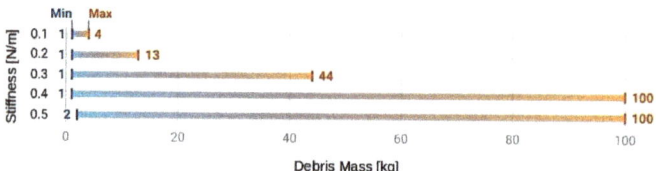

Figure A1. Range of space debris to be targeted for 4 s contact time.

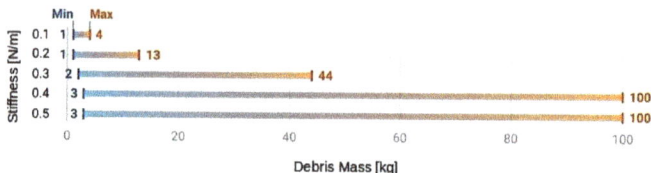

Figure A2. Range of space debris to be targeted for 6 s contact time.

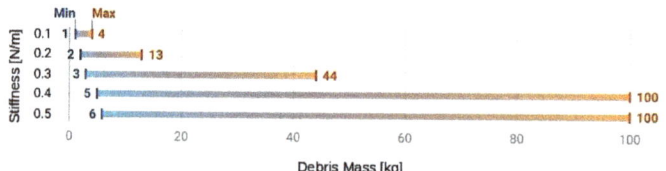

Figure A3. Range of space debris to be targeted for 8 s contact time.

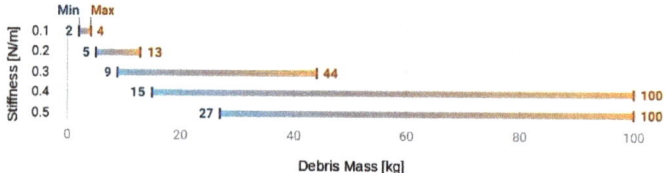

Figure A4. Range of space debris to be targeted for 12 s contact time.

References

1. Bernhard, P.; Deschamps, M.; Zaccour, G. Large satellite constellations and space debris: Exploratory analysis of strategic management of the space commons. *Eur. J. Oper. Res.* **2023**, *304*, 1140–1157. [CrossRef]
2. Space Environment Statistics. Available online: https://sdup.esoc.esa.int/discosweb/statistics/ (accessed on 15 September 2022).
3. Drmola, J.; Hubik, T. Kessler Syndrome: System Dynamics Model. *Space Policy* **2018**, *44–45*, 29–39. [CrossRef]
4. Liou, J.C. Active Debris Removal and the Challenges for Environment Remediation. In Proceedings of the 28th International Symposium on Space Technology, Ginowan, Japan, 29 August–2 September 2012.
5. Aslanov, V.; Ledkov, A. 2—Space Debris Problem. In *Attitude Dynamics and Control of Space Debris During Ion Beam Transportation*; Springer: Berlin, Germany, 2023. [CrossRef]
6. Bonnal, C. Active debris removal Recent progress and current trends. *Acta Astronaut.* **2013**, *85*, 51–60. [CrossRef]
7. Borelli, G.; Gaias, G.; Colombo, C. Rendezvous and proximity operations design of an active debris removal service to a large constellation fleet. *Acta Astronaut.* **2023**, *205*, 33–46. [CrossRef]
8. Færgestad, R.; Holmen, J.; Berstad, T.; Cardone, T.; Ford, K.; Børvik, T. Coupled finite element-discrete element method (FEM/DEM) for modelling hypervelocity impacts. *Acta Astronaut.* **2023**, *203*, 296–307. [CrossRef]
9. Yalçın, B.C.; Martinez, C.; Delisle, M.H.; Rodriguez, G.; Zheng, J.; Olivares-Mendez, M. ET-Class: An Energy Transfer-Based Classification of Space Debris Removal Methods and Missions. *Front. Space Technol.* **2022**, *3*, 23. [CrossRef]
10. Liu, J.; Cui, N.; Shen, F.; Rong, S. Dynamics of Robotic Geostationary orbit Restorer system during deorbiting. *IEEE Aerosp. Electron. Syst. Mag.* **2014**, *29*, 36–42. [CrossRef]

11. Fang, G.; Zhang, Y.; Sun, Y.; Huang, P. On the Stiffness Selection for Tethered Space Robot. In Proceedings of the 2022 IEEE International Conference on Robotics and Biomimetics (ROBIO), Xishuangbanna, China, 5–9 December 2022; pp. 297–302. [CrossRef]
12. Wu, S.; Mou, F.; Liu, Q.; Cheng, J. Contact dynamics and control of a space robot capturing a tumbling object. *Acta Astronaut.* **2018**, *151*, 532–542. [CrossRef]
13. Nishida, S.I.; Uenaka, D.; Matsumoto, R.; Nakatani, S. Lightweight Robot Arm for Capturing Large Space Debris. *J. Electr. Eng.* **2018**, *6*, 271–280. [CrossRef]
14. Somov, Y.; Butyrin, S.; Somov, S. Guidance and Control of a Space Robot-manipulator at Approach and Capturing a Passive Satellite. *IFAC-PapersOnLine* **2019**, *52*, 538–543. [CrossRef]
15. Yan, L.; Xu, W.; Hu, Z.; Liang, B. Multi-objective configuration optimization for coordinated capture of dual-arm space robot. *Acta Astronaut.* **2020**, *167*, 189–200. [CrossRef]
16. Liu, Z.; Lin, T.; Wang, H.; Yue, C.; Cao, X. Design and Demonstration for an Air-bearing-based Space Robot Testbed. In Proceedings of the 2022 IEEE International Conference on Robotics and Biomimetics (ROBIO), Samui, Thailand, 4–9 December 2022; pp. 321–326. [CrossRef]
17. Wang, X.; Shi, L.; Katupitiya, J. A strategy to decelerate and capture a spinning object by a dual-arm space robot. *Aerosp. Sci. Technol.* **2021**, *113*, 106682. [CrossRef]
18. Chen, G.; Wang, Y.; Wang, Y.; Liang, J.; Zhang, L.; Pan, G. Detumbling strategy based on friction control of dual-arm space robot for capturing tumbling target. *Chin. J. Aeronaut.* **2020**, *33*, 1093–1106. [CrossRef]
19. Cao, Y.; Wang, S.; Zheng, X.; Ma, W.; Xie, X.; Liu, L. Reinforcement learning with prior policy guidance for motion planning of dual-arm free-floating space robot. *Aerosp. Sci. Technol.* **2023**, *136*, 108098. [CrossRef]
20. Sze, H.Y.; Chhabra, R. Trajectory Generation for Space Manipulators Capturing Moving Targets Using Transfer Learning. In Proceedings of the 2023 IEEE Aerospace Conference, Big Sky, MT, USA, 4–11 March 2023. [CrossRef]
21. Psomiadis, E.; Papadopoulos, E. Model-Based/Model Predictive Control Design for Free Floating Space Manipulator Systems. In Proceedings of the 2022 30th Mediterranean Conference on Control and Automation (MED), Athens, Greece, 28 June–1 July 2022; pp. 847–852. [CrossRef]
22. Fauré, M.; Henry, D.; Cieslak, J.; Colmenarejo, P.; Ankersen, F. A H∞ control solution for space debris removal missions using robotic arms: The ESA e.Deorbit case. In Proceedings of the 2022 UKACC 13th International Conference on Control (CONTROL), Plymouth, UK, 20–22 April 2022; pp. 122–129. [CrossRef]
23. Kayastha, S.; Katupitiya, J.; Pearce, G.; Rao, A. Comparative study of post-impact motion control of a flexible arm space robot. *Eur. J. Control* **2023**, *69*, 100738. [CrossRef]
24. Viscuso, S.; Gualandris, S.; De Ceglia, G.; Visentin, V. Chapter 18—Shape memory alloys for space applications. In *Shape Memory Alloy Engineering*, 2nd ed.; Concilio, A., Antonucci, V., Auricchio, F., Lecce, L., Sacco, E., Eds.; Butterworth-Heinemann: Boston, MA, USA, 2021; pp. 609–623. [CrossRef]
25. Sinatra, N.R.; Teeple, C.B.; Vogt, D.M.; Parker, K.K.; Gruber, D.F.; Wood, R.J. Ultragentle manipulation of delicate structures using a soft robotic gripper. *Sci. Robot.* **2019**, *4*, eaax5425. [CrossRef]
26. Li, X.; Chen, Z.; Wang, Y. Detumbling a Space Target Using Soft Robotic Manipulators. In Proceedings of the 2022 IEEE International Conference on Mechatronics and Automation (ICMA), Guilin, China, 7–10 August 2022; pp. 1807–1812. [CrossRef]
27. Wang, W.; Tang, Y.; Li, C Controlling bending deformation of a shape memory alloy-based soft planar gripper to grip deformable objects. *Int. J. Mech. Sci.* **2021**, *193*, 106181. [CrossRef]
28. Lu, Y.; Xie, Z.; Wang, J.; Yue, H.; Wu, M.; Liu, Y. A novel design of a parallel gripper actuated by a large-stroke shape memory alloy actuator. *Int. J. Mech. Sci.* **2019**, *159*, 74–80. [CrossRef]
29. Feng, L.; Martinez, P.; Dropmann, M.; Ehresmann, M.; Ginsberg, S.; Herdrich, G.; Laufer, R. MEDUSA—Mechanism for Entrapment of Debris Using Shape memory Alloy. In Proceedings of the 1st ESA NEO and Debris Detection Conference, Darmstadt, Germany, 22–24 January 2019.
30. Jiang, H.; Hawkes, E.W.; Fuller, C.; Estrada, M.A.; Suresh, S.A.; Abcouwer, N.; Han, A.K.; Wang, S.; Ploch, C.J.; Parness, A.; et al. A robotic device using gecko-inspired adhesives can grasp and manipulate large objects in microgravity. *Sci. Robot.* **2017**, *2*, eaan4545. [CrossRef] [PubMed]
31. Hashizume, J.; Huh, T.M.; Suresh, S.A.; Cutkosky, M.R. Capacitive Sensing for a Gripper with Gecko-Inspired Adhesive Film. *IEEE Robot. Autom. Lett.* **2019**, *4*, 677–683. [CrossRef]
32. Estrada, M.A.; Hockman, B.; Bylard, A.; Hawkes, E.W.; Cutkosky, M.R.; Pavone, M. Free-Flyer Acquisition of Spinning Objects with Gecko-Inspired Adhesives. In Proceedings of the 2016 IEEE International Conference on Robotics and Automation (ICRA), Stockholm, Sweden, 16–21 May 2016. [CrossRef]
33. Su, Y.; Hou, X.; Li, L.; Cao, G.; Chen, X.; Jin, T.; Jiang, S.; Li, M. Study on impact energy absorption and adhesion of biomimetic buffer system for space robots. *Adv. Space Res.* **2020**, *65*, 1353–1366. [CrossRef]
34. Mitros, Z.; Rekleitis, G.; Papadopoulos, E. Impedance control design for on-orbit docking using an analytical and experimental approach. In Proceedings of the 2017 25th Mediterranean Conference on Control and Automation (MED), Valletta, Malta, 3–6 July 2017; pp. 1244–1249. [CrossRef]
35. Zhang, G.; Zhang, Q.; Feng, Z.; Chen, Q.; Yang, T. Dynamic modeling and simulation of a novel mechanism for adhesive capture of space debris. *Adv. Space Res.* **2021**, *68*, 3859–3874. [CrossRef]

36. Moosavian, S.A.A.; Papadopoulos, E. Multiple impedance control for object manipulation. In Proceedings of the 1998 IEEE/RSJ International Conference on Intelligent Robots and Systems Innovations in Theory, Practice and Applications (Cat. No. 98CH36190), Victoria, BC, Canada, 13–17 October 1998; Volume 1, pp. 461–466.
37. Hogan, N. Impedance control: An approach to manipulation. In Proceedings of the 1984 American Control Conference IEEE, San Diego, CA, USA, 6–8 June 1984; pp. 304–313.
38. Koga, K.; Fukui, Y. Deorbiting of Satellites by a Free-Flying Space Robot by Combining Positioning Control and Impedance Control. In Proceedings of the 2022 22nd International Conference on Control, Automation and Systems (ICCAS), Jeju, Republic of Korea, 27 November–1 December 2022; pp. 965–971. [CrossRef]
39. ClearSpace. Available online: https://clearspace.today/ (accessed on 15 September 2022).
40. Biesbroek, R.; Aziz, S.; Wolahan, A.; Cipolla, S.; Richard-Noca, M.; Piguet, L. The clearspace-1 mission: ESA and clearspace team up to remove debris. In Proceedings of the European Conference on Space Debris (Virtual), Darmstadt, Germany, 20–23 April 2021.
41. DiscoWeb. Available online: https://discosweb.esoc.esa.int/objects (accessed on 15 September 2022).
42. ESA Space Debris Mitigation WG. *ESA Space Debris Mitigation Compliance Verification Guidelines*; European Space Agency: Paris, France, 2015.
43. Hawkes, E.W.; Jiang, H.; Cutkosky, M.R. Three-dimensional dynamic surface grasping with dry adhesion. *Int. J. Robot. Res.* **2015**, *35*, 943–958. [CrossRef]
44. Suresh, S. Engineering Gecko-Inspired Adhesives. 2020. Available online: https://phowpublished.stanford.edu/cp134gr3166 (accessed on 15 September 2022).
45. Fehse, W. *Automated Rendezvous and Docking of Spacecraft*; Cambridge University Press: Cambridge, UK, 2003; Volume 16.
46. Brannan, J.; Scott, N.; Carignan, C. Robot Servicer Interaction with a Satellite During Capture. In Proceedings of the International Symposium on Artificial Intelligence, Robotics and Automation in Space (iSAIRAS), Madrid, Spain, 4–6 June 2018.
47. Stronge, W.J. *Impact Mechanics*; Cambridge University Press: Cambridge, UK, 2018.
48. Mitros, Z.; Paraskevas, I.S.; Papadopoulos, E.G. On robotic impact docking for on orbit servicing. In Proceedings of the 2016 24th Mediterranean Conference on Control and Automation (MED) IEEE, Athens, Greece, 21–24 June 2016; pp. 1120–1125.
49. Nanos, K.; Xydi-Chrysafi, F.; Papadopoulos, E. On impact de-orbiting for satellites using a prescribed impedance behavior. In Proceedings of the 2019 IEEE 58th Conference on Decision and Control (CDC) IEEE, Nice, France, 11–13 December 2019; pp. 2126–2131.

Disclaimer/Publisher's Note: The statements, opinions and data contained in all publications are solely those of the individual author(s) and contributor(s) and not of MDPI and/or the editor(s). MDPI and/or the editor(s) disclaim responsibility for any injury to people or property resulting from any ideas, methods, instructions or products referred to in the content.

Article

Comparison between Different Re-Entry Technologies for Debris Mitigation in LEO

Francesco Barato

Department of Industrial Engineering, Università degli Studi di Padova, Via Venezia 1, 35121 Padova, Italy; francesco.barato@unipd.it

Abstract: The population of satellites in Low Earth Orbit is predicted to growth exponentially in the next decade due to the proliferation of small-sat constellations. Consequently, the probability of collision is expected to increase dramatically, possibly leading to a potential Kessler syndrome situation. It is therefore necessary to strengthen all the technologies required for collision avoidance and end-of-life disposal of new satellites, together with active debris removal of current and potential future dead satellites. Both situations require the lowering of the altitude of a satellite up to re-entry. In this paper several de-orbiting technologies are evaluated: natural decay, chemical propulsion (solid and liquid), electric propulsion, drag sail, electrodynamic tether, and combinations of the previous ones. The comparison considers the initial altitude, system mass, de-orbiting time, collision probability during descent, reliability, and technological limits. Differences between active debris removal and satellite end-of-life self-disposal are taken into account. Moreover, the different types of re-entry, controlled vs. non-controlled, expendable vs. reusable system, demisable vs. non-demisable system are also discussed. Finally, the possibility to operate the satellite in Very Low Earth Orbits with a propulsion system for drag compensation and passive re-entry at end of life is investigated.

Keywords: de-orbiting; end-of-life disposal; space debris; propulsion; tether; sail; orbital decay; LEO; VLEO; drag compensation

1. Introduction

In recent times there has been an increasing concern about debris in Low Earth Orbit (LEO); in fact it is more and more frequent to find in the space-related news episodes of collision avoidance or sometimes even real (deliberate or unwanted) collisions or fragmentations of space objects [1,2]. This is due to the fact that the satellite population in LEO is growing exponentially as the number of satellites put into orbit is much greater than quantity of the ones that are removed. Moreover, actual predictions of future space trends show a further order of magnitude increase due to the rise of the so-called mega-constellations [3,4]. Space experts, shareholders and institutions all around the world have often claimed that the current behavior cannot be tolerated as it is in the future, otherwise the possibility of Kessler syndrome with its consequent terrible effects could become real [5,6]. Even without a catastrophic scenario, a crowded space environment could render it more difficult to operate in space, as, for example, a significant number of collision avoidance maneuvers could impact the delta-v budget of the mission and the cost of operations. Therefore, it is necessary to take actions in order to limit as much as possible the amount of inoperative material orbiting in LEO. At these altitudes, the common way to clean the orbits is trough deorbiting and re-entry/disintegration of the space object.

It is worth noting that the largest portion (around 2/3) of orbital debris is concentrated in LEO, and that only 6% of Earth orbiting objects are operational payloads [7,8]. Moreover, LEO altitude distribution shows a peak around 800 km, which is in fact one of the favorite orbital altitudes and the target of the new small-sat rideshare launch missions like the recent Transporter service by SpaceX [9].

There are plenty of important aspects related with space debris like monitoring, prediction, traffic management, protection, policy, autonomous detection, operations and so forth [10–14]. This paper will focus only on a single, unavoidable aspect, the means for deorbiting satellites at the end of life. This topic has been already discussed previously [15–18], and many numerical methods for the solution of the perturbed two-body problem have been developed, e.g., [19–21]. However, here some new further-simplified analyses will be shown, with a specific focus on the system level impact of the added devices, in particular total mass and deorbiting time.

Generally, the principal proposed ways to deorbit a satellite are the natural aerodynamic decay (sometimes with a drag-boosting device, i.e., a drag sail), propulsive de-orbit, electrodynamic tethers, solar sails.

Compared to previous works [15–18], the range of the initial conditions for the comparison (altitude, decay time, inclination, duty cycle, deorbiting profile etc.) has been extended. Moreover, the combinations between different technologies will be also discussed. Finally, the paradigm of drag compensation will be directly compared with conventional deorbit. Other aspects closely related with deorbiting will be also reviewed.

The aim of this work is to better understand the different available options with their strength and weaknesses and provide simple tools, guidelines and warnings in order to support future selections or developments.

2. Deorbiting Technologies

The simplest way to deorbit a satellite in LEO is using the natural aerodynamic decay. This is a passive and inexpensive solution. However, the drag D produced is linearly dependent on the atmospheric density ρ:

$$D = 0.5 \rho v^2 c_d A \qquad (1)$$

The drag coefficient c_d is considered near 2 for a typical satellite (2.2 used in this paper), where A is the frontal area respect to the incoming flow (the opposite of the velocity vector). Here the density has been calculated with the US Standard Atmosphere 1976 model, USSA1976 [22]. This model does not take into account temporal and horizontal spatial variations, but only the average behavior with altitude. More complete models are available [23], and it is important to remember that the actual density at a certain altitude can change significantly due to variations in Sun activity. However, once this caveat is known, the current model anyway is considered sufficient to be used for the relative comparison of the different deorbiting technologies.

The atmospheric density has roughly an exponential behavior with altitude. Therefore, the lifetime of a dead satellite is also an exponential function of the altitude. Current legislations foresee a decay time below 25 years for non-operational satellites. It is possible to see that the corresponding limit altitude is around 600 km. It is worth remembering that the peak of debris and the current favorite altitude is slightly above this value. However, from the current analyses and trends it is possible to predict (and also to encourage) that sooner or later the legislation should become stricter, reducing the time allowed to complete deorbit. In this case the limit altitude will decrease significantly. The author proposes the possibility to put a mandatory fee on deorbit, with the fee proportional to the deorbit time, or time spent in orbit after end of life in case of failed deorbit (forcing an active debris removal in the latter).

In order to improve the decay time, it is possible to add a drag sail [24–28]. This device is deployed at the end of life to drastically increase the frontal area and consequently the drag, therefore reducing the deorbit time. The device needs an actuation system, i.e., a mechanism to deploy the sail correctly and a communication link or an internal computer that starts the procedure when needed. Consequently, at least two failure points are introduced.

The mass of the drag sail is approximated as a linear function of its area. Typical values of area density are around 75 g/m^2 with up to dozens of m^2 of area [17]. After deployment, the satellite will theoretically passively deorbit. However, in the general case the attitude of the satellite could change with time and the drag sail could not present its full area against the flow, reducing its efficacy. Three possibilities are then possible: to accept a performance degradation, but this choice is not preferable as it comes with a significant uncertainty; to add an attitude control during re-entry, perhaps using one already present in the satellite (which, however should cope with different forces compared to its original design requirements); or, finally, find a suitable sail shape design that would preserve its positional stability with respect to the incoming flow. The last possibility seems the most promising, even if not fully developed.

However, the drag sail linearly decreases the decay time as the inverse of its frontal area. Therefore, it performs very well at low altitudes, but rapidly loses effectiveness at higher altitudes.

A similar solution is the solar sail. The solar sail is a proposed solution to impart a delta-v on a spacecraft without consuming propellant, but rather using the force from solar radiation. Very interesting for certain missions, it can be potentially used also for deorbiting a spacecraft. The solar pressure is:

$$p_{rad} = I/c \qquad (2)$$

where I is the solar irradiance and c the speed of light [8]. The value of the solar pressure is rather low, around 4.5 µN/m². The altitude at which the solar and aerodynamic pressure are equal is around 600 km. At this altitude, the drag sail is already not the best-performing solution, thus the solar sail is not well suited for deorbit, particularly taking into account that the solar vector is not aligned with the opposite of the velocity vector for most of the orbit (unlike aerodynamic drag), thereby dramatically reducing its effectiveness and requiring an active attitude control to properly orient the sail.

The classical way to deorbit a satellite is to use a propulsion system. The parameters characterizing the propulsion system are the specific impulse Isp, thrust T, total impulse I_{tot}, propellant mass m_p. The propellant mass is dependent on the delta-v (Δv) required for deorbiting trough the Tsiolkovsky equation:

$$\Delta v = Isp \, g_0 \ln(m_i/m_f) \qquad (3)$$

$$m_p = m_i - m_f \qquad (4)$$

where the subscripts i and f stand for initial and final, respectively. For high-thrust systems, an impulsive maneuver can be considered and a Hohmann transfer from the original orbit to another at lower altitude with corresponding lower lifetime can be performed.

$$\Delta v_H = (v - v_a) + (v_p - v_0) \qquad (5)$$

where a stands for apogee, p for perigee, H for Hohmann and v is the instantaneous orbital velocity:

$$v = \sqrt{\mu\left(\frac{2}{r} - \frac{1}{a}\right)} \qquad (6)$$

Which is constant for a circular orbit. μ is the Earth gravitational constant, r the local orbit radius and a is the semi-major axis. Required delta-v and thrust are generally sufficiently high to force the use of a chemical propulsion system, which are divided between solids, liquids and, less frequently, hybrids.

Solid systems have the advantage to be very compact and simple, favoring a dedicated one-shot device [29–31]. Metallized solid propellants release solid particles, so less energetic particle-free formulations are preferable, particularly for deorbit from high altitudes. Solid rockets have relatively high thrusts, which can be an issue for attitude control; in fact, often the motor requires an active thrust vector control, or the spacecraft needs to be spun during firing. Solid rockets are not suited for multifunctional use. On the contrary, liquids tend to be more complex and bulkier, but can be integrated in order to provide other propulsive functions other than deorbiting (like orbit raising and station keeping). Thrust is relatively low so attitude (and trajectory) control is much easier and precise. Hybrid propulsion has its own peculiarities as described in [32–35] and has been proposed for active deorbiting of large items [36,37].

For low-thrust systems, a continuous thrusting phase is necessary. The delta-v is the difference between the original (marked with 0) orbital velocity and the final one.

$$\Delta v = v_0 - v \tag{7}$$

In LEO, the orbits always have a small altitude compared with the Earth radius and consequently a very low eccentricity. Therefore, the difference between a Hohmann transfer and a continuous transfer becomes negligible. The ratio between the continuous delta-v and the impulsive (Hohmann) one is given by the following equation [38]:

$$\frac{\Delta v}{\Delta v_H} = \left[\sqrt{2\left(1 + \frac{2\sqrt{r/r_0}}{r/r_0 + 1}\right)} - 1\right]^{-1} \tag{8}$$

The ratio is only 0.32% higher than one for a deorbiting from 2000 to 300 km (828.2 m/s vs. 825.6 m/s). It is worth noting that, for altitudes above 1400 km, it is more efficient to move the satellite into a disposal orbit slightly above 2000 km than deorbit it [11].

However, thanks to the exponential behavior of the atmospheric density, instead of lowering all the orbit to a new altitude, it is possible to decrease only the perigee to an altitude slightly below the one necessary for a circular orbit to obtain the same decay time. The first half of a Hohmann transfer is roughly one-half of the total delta-v. The delta-v for this option is slightly above one half of the complete Hohmann for the same decay time. This option is possible only for a high-thrust system that can perform an impulsive maneuver. Low-thrust systems are generally of electric type, require power and are consequently limited in thrust by it. The maneuver time is generally non-negligible or even comparable with the decay time, and in fact the two are superposed. The advantage of the electric system is a much higher specific impulse that can provide an attractive mass saving for high delta-v.

Finally, the last option is the electrodynamic tether [17,39–41]. In this case a tether is deployed for re-entry. Electrodynamic tethers collect ionospheric electrons from the plasma environment and re-emit them through a cathode or a "Low-Work-Function" segment of the same tether by using thermionic and photoelectric effects. In both configurations, the resulting electric current flowing through the conductive tether generates a drag Lorentz force thanks to interaction with the Earth's magnetic field. The drag force is rather small, and the deorbiting behavior is very similar to that of an electric thruster, but without propellant consumption. The technology is still not as mature as the classical propulsion; concerns arise regarding the oscillation stability of the tether–satellite system, the probability/consequences of debris impact on the tether and other engineering issues.

The Lorentz drag force F_{et} is described by the following equation:

$$F_{et} = m\frac{\sigma}{\rho}\frac{E_m^2 i_{av}}{v} = m\frac{\sigma}{\rho}vB_m^2\cos^2(i)i_{av} \tag{9}$$

$$E_m = vB_m\cos(i) \tag{10}$$

where B_m is the geomagnetic field, E_m the motional electric field, i_{av} the dimensionless averaged current along the tether, m the tether mass, ρ its density, σ is the tether conductivity. Typical values are ρ = 2700 kg/m^3, I = 3.54 × 10^7 Ω^{-1}m^{-1}, B_m = 0.3 Gauss, i_{av} = 0.25. For a fixed technology, the main parameters affecting the Lorentz drag force are the area and thickness of the tether and the orbital inclination i. The tether becomes less effective approaching polar orbits due to the cos^2 dependency.

The average TRL of the tether systems is the lowest between all the possible options, anyway some devices are already sold as COTS on the market [42–44]. However, performance from datasheet seems to indicate a much higher relative auxiliary mass compared with literature projections. This is probably due to the small size of the devices and or to some compromises necessary to field such a new technology nowadays. The tether can be used also in reverse mode as a propulsion system for orbit raising (requiring power from the spacecraft), however the TRL of the technology is even lower in this case.

The chemical thruster is an active system that requires the satellite to be effective only for a limited time after the start of the disposal maneuver, generally no more than several hours in case the apogee burn is split in several smaller impulses. On the contrary, both the electric thrusters and the tethers need to work reliably for a long time, meanwhile requiring the satellite to control its attitude and basic functions, increasing the probability of failure.

Comparing the forces produced by the various technologies for 10 kg of device mass we obtain the results presented in the following Figure 1. The thrust of a chemical propulsion system is not represented as it is out of scale and not a particular limiting factor. For the electric propulsion only the fixed mass of the thruster is considered, i.e., the propellant mass is excluded as it depends on the specific altitude change. For the electric thruster the mass is computed as a linear function of the required power ($m_t = \alpha P$, $P = 0.5T\, Isp\, g_0/\eta$), considering $\alpha = 20$ kg/kW and a thruster efficiency η of 0.65.

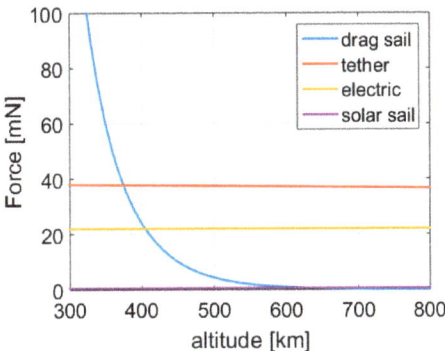

Figure 1. Comparison of different forces produced by different deorbiting devices (10 kg mass) as a function of altitude.

The force of the drag sail is almost an exponential function of the altitude (as the atmospheric density). The thrust of the electric propulsion system is independent of the altitude. The force of the tether is calculated for a 45° orbital inclination, it is rather constant with altitude (slightly decreasing, <10% in LEO) and it is almost double that of the electric thruster. The solar sail force is negligible compared to the other devices (except when the drag sail loses usefulness), thus it is not considered a viable option. Therefore, for the rest of the paper, the term sail will be associated with aerodynamic drag sail. The drag sail is the most effective at very low altitudes while the tether is potentially the most performing solution at the majority of altitudes.

Comparing the tether with the electric propulsion system (as it is the most similar) for different inclinations for 10 kg of device mass, we obtain the results presented in Figure 2. In this case, for the electric propulsion system the propellant mass together with its corresponding structural mass $m = (1 + k)m_p$, $k = 0.12$, have been also included considering a deorbiting from 750 to 200 km. It is possible to see that the superiority of the tether vanishes approaching the polar orbit and remains below the electric thruster for a sun-synchronous orbit, which is, at the moment, one of the most frequent in LEO.

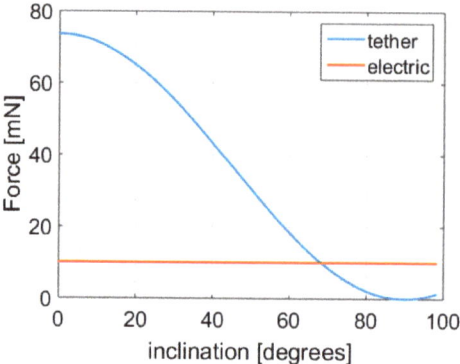

Figure 2. Variation of the tether force with orbital inclination (10 kg mass device).

3. Materials and Methods

In order to compare the different technologies, it is necessary to calculate the deorbiting behavior. For this a simple Hohmann transfer Equation (5) is used. For a half Hohmann transfer only the first term is necessary. Propellant mass is calculated with Equations (3) and (4).

Time for the maneuver can be calculated by simple orbital mechanics [45–47] as half the orbital period T:

$$T = 2\pi \sqrt{a^3/\mu} \tag{11}$$

but is definitely negligible (around 90 min per each full orbit, the number of orbits depending on the number of impulses that composes the total impulse required) with respect to the typical required deorbiting time.

For the calculation of deorbiting time with drag and small forces, the equation of motion is integrated with time:

$$\ddot{\mathbf{r}} + \frac{\mu}{r^3}\mathbf{r} = \mathbf{a}_F \tag{12}$$

where $\mathbf{a}_F$ is the force-induced acceleration (F/m). This is the so-called Cowell approach. It requires a small timestep in order to avoid a large error buildup. The vectorial equation is integrated with an adaptive step 4th–5th-order Runge–Kutta scheme. The error tolerance has been set by repeating the simulation until the relative error between the last two was well below 1%. For the sake of simplicity, the rotation of the atmosphere has been neglected in this work so that the relative velocity is equal to the spacecraft absolute velocity.

Two semi-analytical techniques have been also used to drastically reduce the computational time. In case of an initial circular orbit and a slow spiraling down the following equation has been used:

$$r \approx \frac{r_0}{\left(1 - \frac{F}{m}\frac{t}{v_0}\right)^2} \tag{13}$$

where $t \leq T$ is the time considered for integration. For elliptical orbits the following approach has been used instead. We assume that a is almost constant during one orbit:

$$a = \frac{2a^2}{\mu}dE \rightarrow \Delta a_{rev} = \frac{2a^2}{\mu}\Delta E_{rev} \tag{14}$$

The energy decay during one orbit is equal to the work of the forces applied to the spacecraft:

$$\Delta E_{rev} = \int_0^T F/mv dt \tag{15}$$

Which in case of drag:

$$F/m = -D/m = -0.5\rho v^2/B_c \tag{16}$$

where B_c is the ballistic coefficient:

$$B_c = m/(c_d A) \tag{17}$$

Substituting in Equations (14) and (15):

$$\Delta a_{rev} = -\frac{a^2}{B_c \mu} \int_0^T \rho v^2 v dt \tag{18}$$

The instantaneous velocity is:

$$v^2 = \sqrt{v_r^2 + v_t^2} = \frac{n^2 a^2}{1-e^2}\left(1 + e^2 + 2e\cos\theta\right) \approx n^2 a^2(1 + 2e\cos\theta) \tag{19}$$

where r means radial, t tangential, e is the eccentricity and θ the true anomaly and:

$$n^2 = (\mu/a)^3 \tag{20}$$

As the eccentricity is small the higher terms on e have been neglected, considering:

$$nt = E - e\sin E \rightarrow dt = \frac{dE}{n}(1 - e\cos E) \tag{21}$$

where E is the eccentric anomaly. Substituting in Equation (18):

$$\Delta a_{rev} = -\frac{a^2}{B_c \mu}\int_0^{2\pi} \rho(na)^3(1 + 2e\cos\theta)^{3/2}(1 - e\cos E)\frac{dE}{n} \tag{22}$$

Also considering Equation (20):

$$\Delta a_{rev} = -\frac{a^2}{B_c}\int_0^{2\pi} \rho(1 + 2e\cos\theta)^{3/2}(1 - e\cos E)dE \tag{23}$$

Remembering that:

$$\cos\theta = \frac{\cos E - e}{1 - e\cos E} \tag{24}$$

We obtain:

$$\Delta a_{rev} = -\frac{a^2}{B_c}\int_0^{2\pi} \rho \frac{(1 + e\cos E)^{3/2}}{\sqrt{1 - e\cos E}} dE \tag{25}$$

Using a simple exponential atmospheric model:

$$\rho = \rho_p \exp\left(\frac{h_p - h}{H}\right) \tag{26}$$

where h is the altitude and H the scale height. Remembering that:

$$r = a(1 - e\cos E) \tag{27}$$

$$r_p = a(1 - e) \qquad r_a = a(1 + e) \tag{28}$$

We obtain:

$$\frac{h_p - h}{H} = \frac{r_p - r}{H} = \frac{ae}{H}(\cos E - 1) \tag{29}$$

Substituting in Equation (25):

$$\Delta a_{rev} = -\frac{a^2}{B_c}\rho_p \int_0^{2\pi} exp(bcosE - b)\frac{(1+ecosE)^{3/2}}{\sqrt{1-ecosE}}dE \qquad (30)$$

With:

$$b = ae/H \qquad (31)$$

Again, neglecting the highe-order terms on e:

$$\frac{(1+ecosE)^{3/2}}{\sqrt{1-ecosE}} \approx 1 + 2ecosE \qquad (32)$$

Thus:

$$\Delta a_{rev} = -\frac{a^2}{B_c}\rho_p \int_0^{2\pi} exp(bcosE)exp(-b)(1+2ecosE)dE \qquad (33)$$

Considering the modified Bessel functions of the first kind:

$$I_j(b) = \frac{1}{2\pi}\int_0^{2\pi} cos(jt)exp(bcost)dt \qquad (34)$$

we obtain:

$$\Delta a_{rev} = -2\pi\frac{a^2}{B_c}\rho_p[I_0 + 2eI_1]exp(-b) \qquad (35)$$

Repeating the process same process:

$$\Delta e_{rev} = -2\pi\frac{a}{B_c}\rho_p[I_1 + e/2(I_0 + I_2)]exp(-b) \qquad (36)$$

The accuracy of the whole model is mainly dependent on the density model. As the atmospheric density is not a perfect exponential, the value of H changes with altitude. The value of H can be calculated from the actual density values at two different altitudes. In the following Figure 3, H is plotted as a function of the altitude for three different spacings (i.e., altitude difference) of the sampling points.

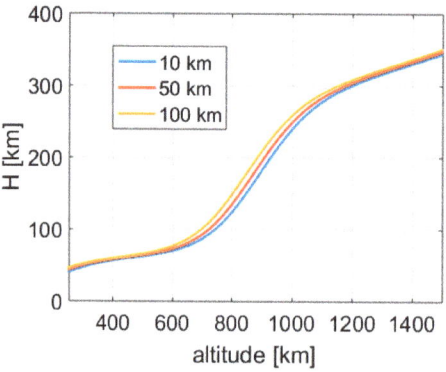

Figure 3. Atmospheric scale factor as a function of altitude, parametric with the distance between sampling points.

The value of H is strongly dependent on the altitude and much less on the spacing between sampling points.

It is important to compare the density calculated with the current model with the original one from the USSA1976 and infer the corresponding error. Figure 4 shows the error for the case of 10 km (a) and 100 km (b) distance between the sampling points. The X-axis represents the distance from the lowest sampling point, while the curves are parametric to the altitude of the lowest sampling point (from 250 to 1500 km). It is possible to see that the exponential density model is obviously correct at the two sampling points (the points 0 and the point + 10 km or +100 km), but it overestimates the density between the two sampling points and underestimate the density above the highest sampling point. As expected, the error is larger at low initial altitudes as the density curve is steeper and the altitude variation is higher as a percentage of the initial altitude.

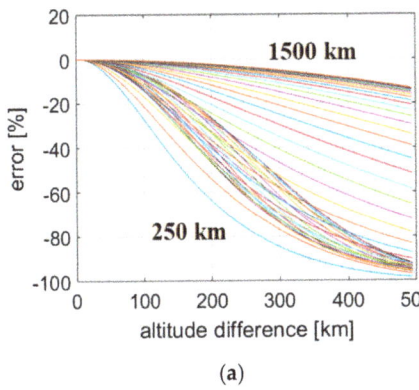
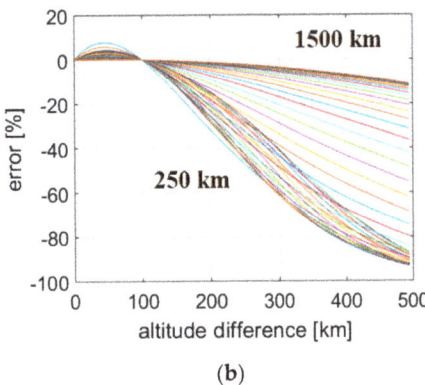

(a) (b)

Figure 4. Error of the exponential density model compared with the USSA1976 for: (a) 10 km sampling spacing; (b) 100 km sampling spacing. Parametric with initial altitude (250 to 1500 km at 25 km steps).

As the model is exponential, a slight error on the curve slope means that the density ratio between the estimated value and the actual one approach zero and the error asymptotically goes to -100%. Even if this error could seem apparently high, for our proposes it is important to underline the following aspect. To determine the decay of the spacecraft, it is important to correctly estimate the energy lost due to the atmospheric drag. Therefore, the accuracy near the perigee, where the drag is high, is very important, while the need for accuracy at the apogee is almost negligible.

For this reason, if someone sets a very small sampling distance, the density calculated with the simplified model will be almost always underestimate along the orbit, and the predicted decay time will be longer, but the error on the decay time will still be reasonable as the error is very small near the perigee. If the distance of the sampling points is increased, the overestimation of the density near the perigee will decrease the decay time, while the underestimation at high altitude will still increase it. After a series of trial and errors, it has been assessed that a sampling distance around 50 km gives generally the best results (Figure 5), with the two effects almost cancelling each other and the decay time prediction becoming precise in the order of an error of several percent.

For example, as shown in Figure 6, for the natural decay of a 50 kg satellite with 400 kg/m^3 density from an initial elliptical orbit with 275 km perigee and 700 km apogee, the predicted deorbit time with the integration of the equations of motion is 347 days, while with the approximated model (50 km sampling distance) is 349 days (0.6%). Such an accurate prediction comes with a dramatically reduction in computational time.

Figure 7 shows that the variation of the semi-major axis during one orbit is negligible and that the eccentricity is very low, justifying the corresponding hypotheses. It is also worth noting from Figure 8 that the flight-path angle is very small in LEO, as any elliptical orbit has a very low eccentricity anyway.

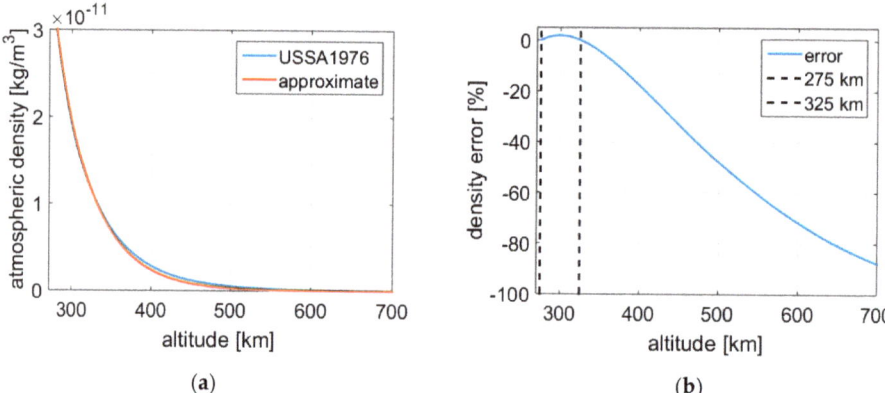

Figure 5. Comparison between the USSA1976 density model and approximate exponential model: (**a**) density vs. altitude; (**b**) error with altitude of the exponential model for a 50 km sampling spacing.

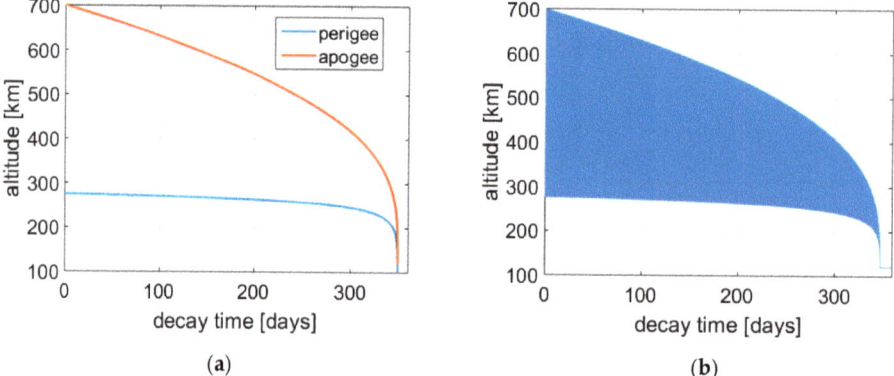

Figure 6. Natural orbital decay from an initial elliptical orbit (275 × 700 km): (**a**) approximate model; (**b**) numerical integration of the equations of motion.

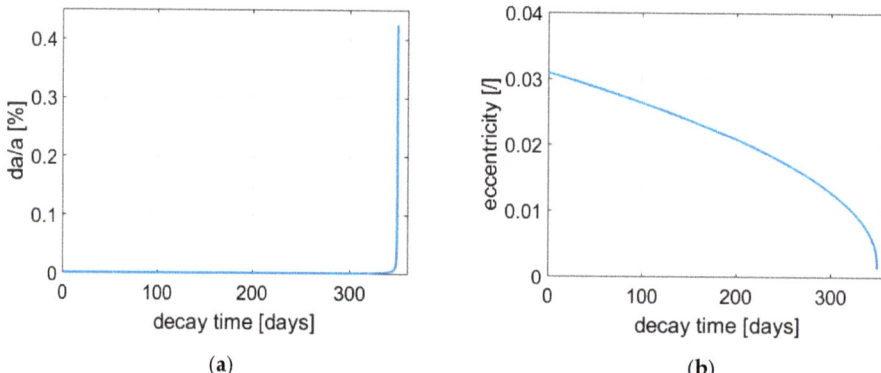

Figure 7. Orbital decay calculated with the approximate model: (**a**) relative variation of the semi-major axis for each orbit; (**b**) eccentricity evolution with time.

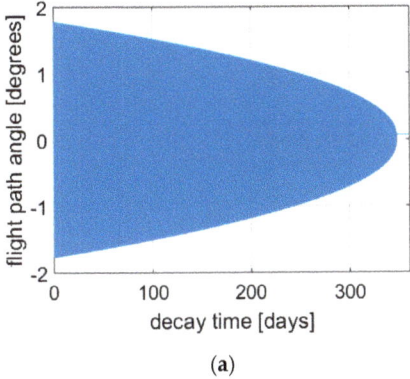 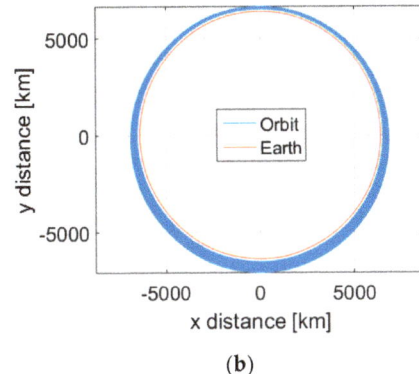

(a)   (b)

Figure 8. Natural orbital decay from an initial elliptical orbit (275 × 700 km) calculated through the numerical integration of the equations of motion: (**a**) flight-path angle; (**b**) orbits' shape.

Consequently, the direction of the spacecraft velocity is almost perfectly orthogonal to the instantaneous orbit radius even far from the apsides and, therefore, gravitational losses are negligible even for a continuous thrusting system.

4. Results

With the simple numerical tools described in the previous chapter it is now possible to determine the specific behavior of the different deorbiting technologies.

4.1. Natural Decay

First of all, as a reference, it is important to determine the natural decay of a typical satellite into LEO without the use of a specific device.

The decay time mainly follows the (inverse) behavior of density with altitude, so it is exponentially longer as the altitude increases (Figure 9). Moreover, the natural decay accelerates as the altitudes decreases, so the majority of the time is spent near the original altitude and the majority of the fall happens near the end of the decay time (Figure 10). This means also that in the case of deorbiting with a special device like a chemical thruster, the majority of the reduction in decay time is obtained in the initial part of the altitude decrease.

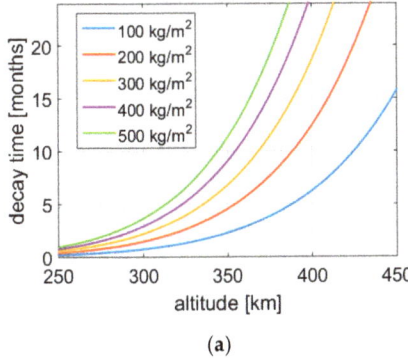 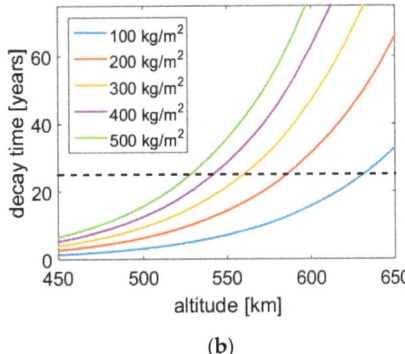

(a)   (b)

Figure 9. Natural decay time as a function of altitude, and parametric with spacecraft mass-to-area ratio: (**a**) lower altitudes; (**b**) higher altitudes, black line is the 25 years limit.

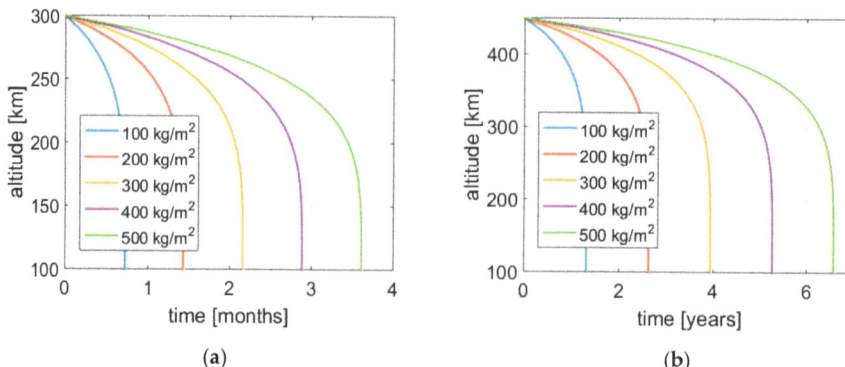

Figure 10. Natural decay of a satellite with time, and parametric with spacecraft mass-to-area ratio: (a) 300 km initial altitude; (b) 450 km initial altitude.

The decay time is also linearly dependent on the ballistic coefficient, which is mainly determined by the ratio between the mass and the frontal area of the satellite. This ratio depends on the scale and the design of the system. For the same, identical satellite (shape and density), the scale should change the ballistic coefficient as the volume to ratio, i.e., a power of 3/2 times the linear size of the system. However, small satellites have typically higher density (around 1000–2000 kg/m^3 for a cubesat) than the larger ones (even below 100 kg/m^3, like the Hubble Space telescope), which mitigates the scale effect. Thus, the majority of satellites have a mass-to-area ratio between 100 and 500 kg/m^2 (0.01 to 0.002 m^2/kg). Anyway, specific designs with large, deployable surfaces can have very low ballistic coefficients, while in the opposite circumstances other designs elongated in the direction of the flow aimed at low-altitude flying (like GOCE [48]) can have a particularly high ballistic coefficient.

From Figure 9, it is possible to see that the 25-year limit is achieved around 500–650 km, while a re-entry in less than five years necessitates an altitude below 300–400 km. Figures 9 and 10 consider circular orbits and the natural decay corresponds to a spiraling down of the satellite trajectory where the eccentricity remains negligible.

The almost exponential trend of the atmospheric density with altitude has a dramatic influence on deorbiting behavior. In the case of an elliptical orbit, the peak of drag at the perigee induces a strong reduction of the apogee altitude in a process of circularization. When the eccentricity becomes very small, the perigee starts to fall as the apogee and the orbit begin spiraling down.

The deorbiting time depends on the orbital energy and the energy dissipation. The first parameter is related to the orbit semi-major axis, while the second is strongly related with the minimum (i.e., perigee) altitude.

In Figure 11 it is possible to see that the same spacecraft has nearly the same decay time (around 2 months) for a more energetic orbit with a lower perigee ($a = R_t + 450$ km, $h_p = 200$ km, $h_a = 700$ km) and a less energetic circular orbit ($h = 316$ km). This is due to the fact that the first, more energetic orbit experiences a larger energy dissipation driven by the higher drag at the (lower) perigee.

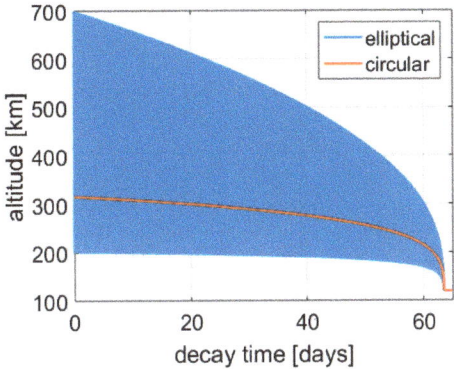

Figure 11. Altitude vs. time for an elliptical orbit (200 × 700 km) and a circular orbit (316 km).

4.2. Drag Sail

Considering the drag sail, it is possible to plot the predicted decay time as a function of the altitude for different ratios of the sail device mass to the spacecraft total mass (Figure 12). The behavior is exactly the same of the natural decay with a simple boost of effectiveness. As already said, due to the exponential behavior of atmospheric density, the decay time becomes too long at higher altitudes. A relatively light system can shift the 25-year limit to over 1000 km. In order to further decrease the decay time, it is necessary to increase the drag sail area, and consequently mass. However, technical feasibility and operational reliability become more and more doubtful. In this case, a chemical propulsion system or a combination of a drag sail with the satellite propulsion system (often already present for other mobility needs) seems to represent a better choice.

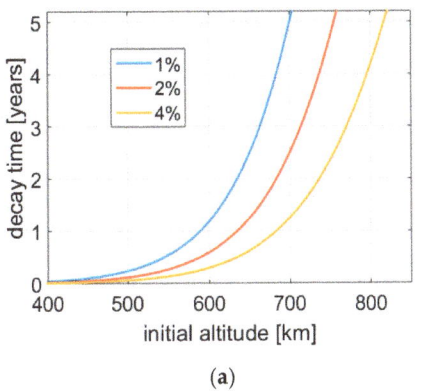

(a)

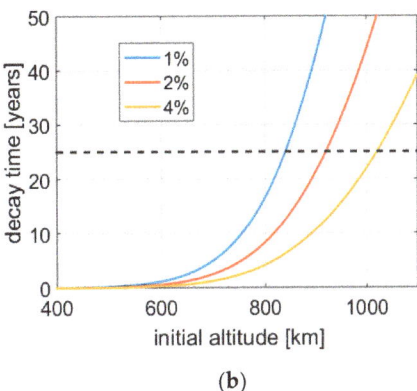

(b)

Figure 12. Decay time as a function of altitude for three different drag sail mass fractions: (a) lower altitudes; (b) higher altitudes, with black line being the 25-year limit.

Sometimes, it is argued that the reduction in the decay time of a drag sail is proportional to its area, and that consequently the probability of impact remains unchanged. This is formally correct, but some aspects should be considered:

1. The impact of small debris with the sail has probably fewer effects and consequences than an impact with the satellite;
2. In case of cooperative systems, a collision avoidance process can occur [49]. This maneuver is based on an uncertainty area of impact that is much larger than the satellite itself and is thus not a linear (but rather sublinear) function of the satellite

area. So, while the probability of passive impact remains constant, the probability of a collision avoidance maneuver by a third party is definitely reduced as the falling time is shortened;

3. It is much easier and cheaper to track a large falling satellite for a shorter time (with the sail) than its original, way smaller, counterpart for a much longer time.

Thus, the drag sail remains an interesting option for deorbiting purposes.

For an elliptical orbit with a low perigee (perhaps as a consequence of an active partial deorbit with a chemical system, as shown in Section 4.7 about combined solutions), we think it is possible that the drag sail could be opened near the perigee to exploit the region of peak drag, and closed in the rest of the orbit where this is little effective. In this way, the probability of impact is widely reduced. However, this solution is probably not worth the added complexity and risks of having an active solar sail be deployed and closed repeatedly once every orbit.

4.3. Chemical Propulsion

With a chemical propulsion system, it is possible to perform a Hohmann transfer from the original orbit to a parking orbit with a defined natural decay time. The mass of the propulsion systems mainly depends on the propellant mass that in turn follows the delta-v between the two orbits. Chemical propulsion has the advantage of being a very fast disposal option, with the Hohmann transfer requiring half an orbit (around 45 min in LEO) for a single impulsive maneuver, or several hours in case the perigee lowering and the final orbit circularization are divided in several burns due to limitation in thrust (and consequently maximum time spent at the apsides).

Figures 13–18 have been computed with $k = 0.12$ and $Isp = 300$ s. For delta-v up to 500 m/s and $Isp = 300$ s, propellant mass is almost linear with delta-v (10% error). Chemical propulsion allows satellites to deorbit from any altitude; however, the required mass can be significant, and easily represent more than 10% of the total satellite mass.

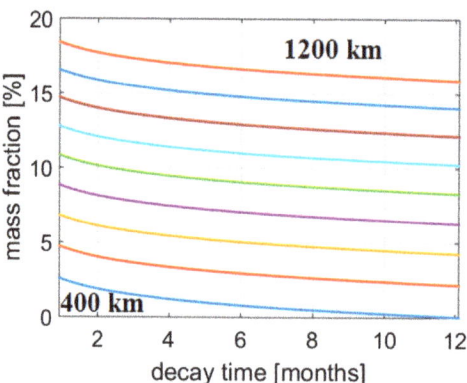

Figure 13. Deorbiting with chemical propulsion with a full Hohmann transfer, parametric with initial altitude (400–1200 km). Propulsion system mass fraction vs. decay time.

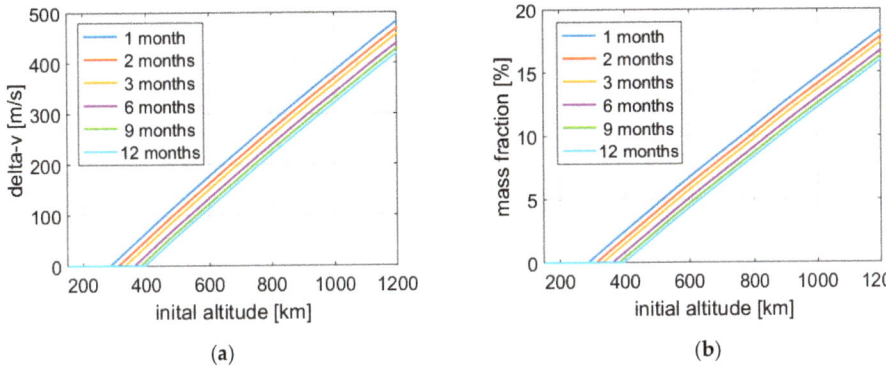

Figure 14. Deorbiting with chemical propulsion with a full Hohmann transfer, parametric with decay time: (**a**) delta-v; (**b**) propulsion system mass fraction.

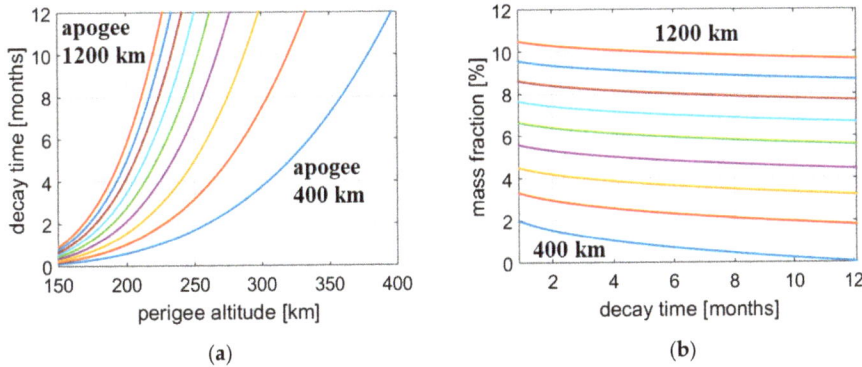

Figure 15. Deorbiting with chemical propulsion with a half Hohmann transfer, parametric with initial altitude (400–1200 km): (**a**) decay time vs. perigee altitude; (**b**) propulsion system mass fraction vs. decay time.

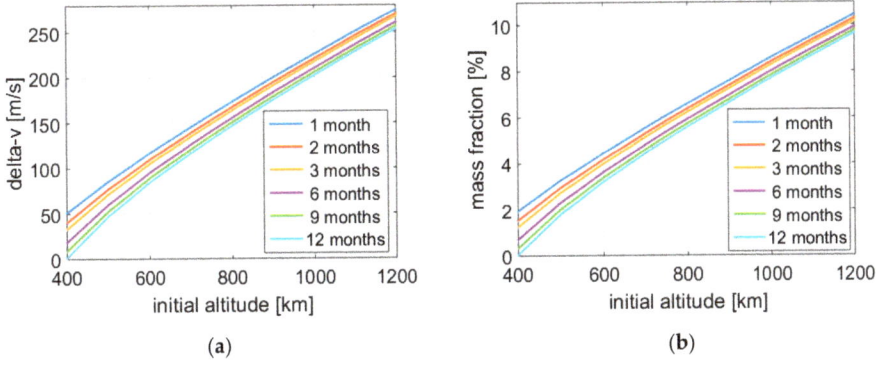

Figure 16. Deorbiting with chemical propulsion with a half Hohmann transfer, parametric with decay time: (**a**) delta-v; (**b**) propulsion system mass fraction.

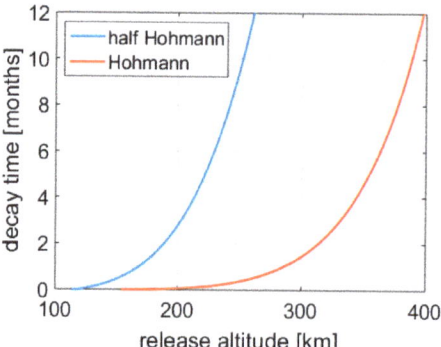

Figure 17. Deorbiting with chemical propulsion, full Hohmann vs. half Hohmann, decay time vs. release altitude; 800 km initial altitude.

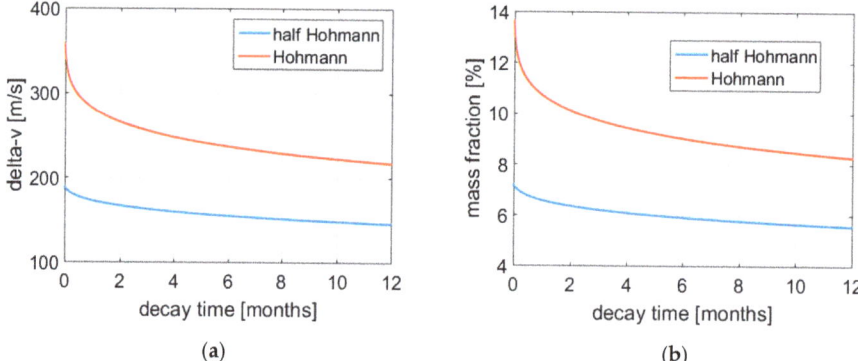

Figure 18. Deorbiting with chemical propulsion, full Hohmann vs. half Hohmann, 800 km initial altitude: (**a**) delta-v; (**b**) propulsion system mass fraction.

In order to reduce this mass, it is possible to increase the altitude of the parking orbit, with the consequence of a longer total deorbiting time. However, as it can be seen in Figures 13 and 14, the system mass has a low sensitivity to the total decay time, as the latter changes tremendously for a rather small variation in the final release altitude, thus not affecting much the propulsion system mass (the cases where the final and initial orbits are near, and where the propulsion system mass is small, are excluded).

As already pointed out earlier, a more efficient possibility is to simply decrease the perigee of the original orbit with a half Hohmann transfer up to a point where the decay time corresponds to the target one (Figures 15 and 16). In this case, contrary to the circular orbit disposal, the decay time is not only dependent on the final (perigee) release altitude but also on the initial one (which remains the apogee altitude), as can be seen in Figure 15a.

Thanks to the exponential behavior of atmospheric density, the perigee of the new orbit is not much lower than the corresponding circular orbit (Figure 17). Consequently, the total delta-v and mass are only slightly higher than half the Hohmann transfer to the circular orbit (around 60%), guaranteeing a significant mass saving. The total mass is thus almost always below 10% of the satellite mass. The qualitative behavior remains the same as before, as it can be seen in Figure 18. The only advantage of the circular orbit disposal is that the satellite is immediately placed far outside the original orbit, while with the more efficient elliptic disposal, the apogee slowly descends from the initial altitude.

The mass model considered for this paper is very simplified. In reality the chemical propulsion system mass is a sublinear function of the propellant mass, particularly for small sizes and lower propellant masses. Thus, if a propulsion system is already onboard for other purposes like station keeping and/or orbit raising, the idea to use it also as a deorbiting device could guarantee important mass and cost benefits, together with the simplicity of having only a single system.

4.4. Electric Propulsion

Similarly to the chemical thruster, an electric thruster can also be used to lower the altitude up to a point where the aerodynamic drag will complete the deorbit. To do so for a fixed total time, it is necessary to increase the thrust of the electric propulsion system in order to speed up the active phase of the descent and let the natural decay do the rest of the deorbit in the remaining time. In this way, the delta-v is reduced and consequently some propellant is saved. However, this requires a larger thruster, and thus is heavier and more expensive. This situation is represented in Figures 19–22.

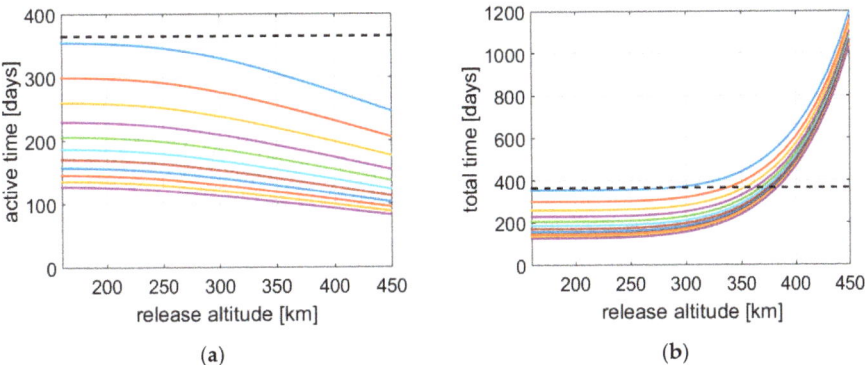

Figure 19. Electric propulsion deorbiting with a combination of an active phase and a passive natural decay, parametric with thrust (0.5–1.5 mN): (**a**) active time vs. release altitude; (**b**) total time vs. release altitude. Black line is the one-year limit.

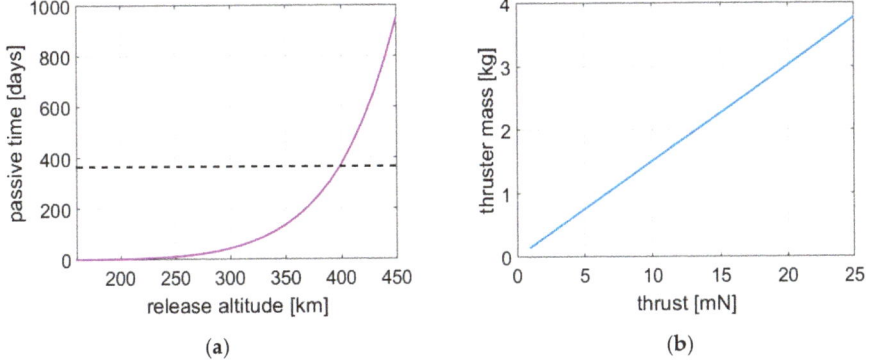

Figure 20. Electric propulsion deorbiting with a combination of an active phase and a passive natural decay: (**a**) passive time vs. release altitude; (**b**) thruster mass vs. thrust.

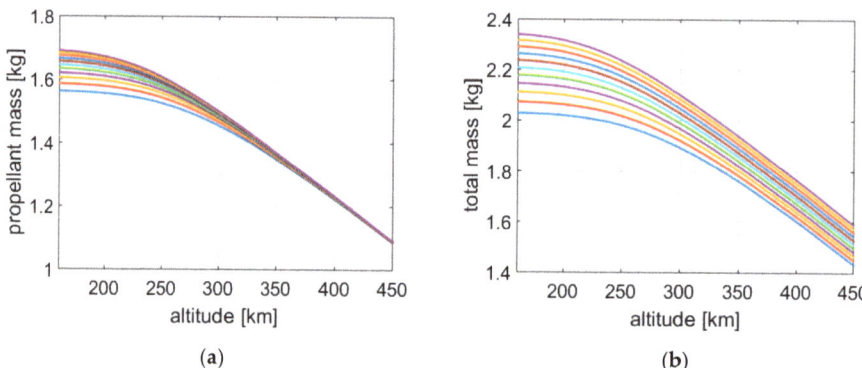

Figure 21. Electric propulsion deorbiting with a combination of an active phase and a passive natural decay, with parametric thrust (0.5–1.5 mN): (**a**) propellant mass vs. release altitude; (**b**) total mass vs. release altitude.

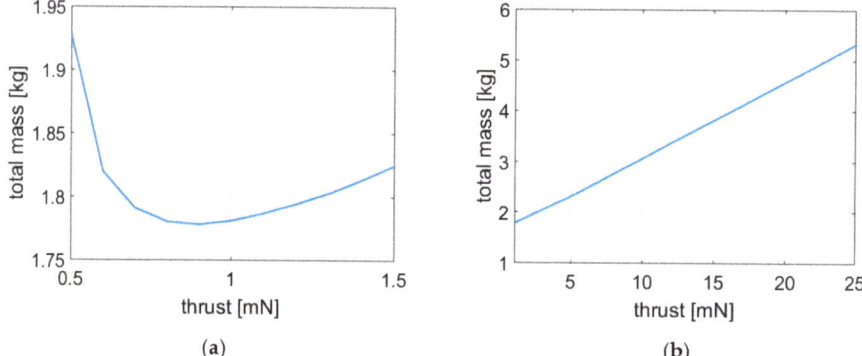

Figure 22. Electric propulsion deorbiting with a combination of an active phase and a passive natural decay, total mass vs. thrust: (**a**) lower thrusts; (**b**) higher thrusts.

A satellite of 50 kg, with a density of 400 kg/m^3, has been considered. The electric propulsion system has an *Isp* of 1000 s, α = 20 kg/kW, η = 0.65 and k = 0.25. The initial altitude is 850 km. The spacecraft is deorbited with the propulsion system up to a release altitude where it is left decaying naturally. The total deorbit time has been fixed in one year. The total time is the sum of the active phase and the passive decay. The passive decay is an exponential function of altitude, see Figure 20a. The release altitude should be less than 400 km in order to be compliant with the one-year limit. The active time is almost linear with the release altitude (Figure 19a), except at low altitudes where the drag force is comparable with the propulsion system thrust. The active time is almost linear with the thrust. The total time increases exponentially with the release altitude but intercepts the same limit at lower altitudes for lower thrusts.

The propellant mass follows the active time behavior with respect to the release altitude, and it is slightly dependent on thrust at low altitudes because of the increased aerodynamic drag savings at lower thrusts. The thruster mass is proportional to the thrust and consequently increases as the active decay time is reduced. The total mass of the propulsion system increases with the thrust and as the release altitude is decreased.

Once a specific total decay time is fixed (in this case one year), it is possible to plot the total mass of the propulsion system as a function of a single parameter, for example the thrust of the electric thruster. Higher thrusts mean shorter active descent time and the possibility to release the satellite at higher altitudes, saving propellant mass. However, as already seen with the chemical system, the exponential behavior of density and decay time with altitude gives a relatively low variation of the release altitude. Moreover, in this case, the total mass behavior seems dominated by the variation of the thruster mass with thrust (Figure 22b) even if a minimum of the curve is present (Figure 22a).

The results are dependent on the choices of thruster efficiency and specific impulse. Higher Isp and lower thruster efficiencies will exacerbate the optimization toward a minimum thrust solution. Higher initial altitudes will do the opposite. It is worth noting that larger thrusters, particularly at small scales, could provide much better performance in terms of Isp and efficiency. This non-linear effect could affect the optimization results toward a higher thrust solution. Technical limits in the total firing time could also have the same consequence.

For an electric propulsion system, it is also possible to activate the thruster only for a fraction of the orbit, around the apogee, in order to lower the perigee down to an altitude where the aerodynamic drag prevails and deorbits the spacecraft. This is analogous of the half Hohmann strategy for the chemical propulsion system, providing a propellant mass saving. However, in this case the deorbit is slow, and, for the same thrust, the decay time gets longer. In Figures 23–26, an example is shown of an electric thruster that is activated in an arc of only 18° before and after the apogee, so 36° in total or 10% of the entire orbit.

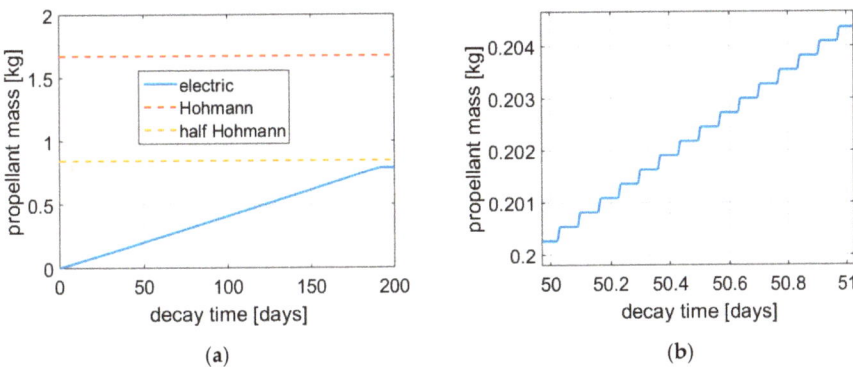

Figure 23. Propellant mass consumption for an electric propulsion system that is activated only +/− 18° around the apogee: (**a**) full picture; (**b**) zoom in.

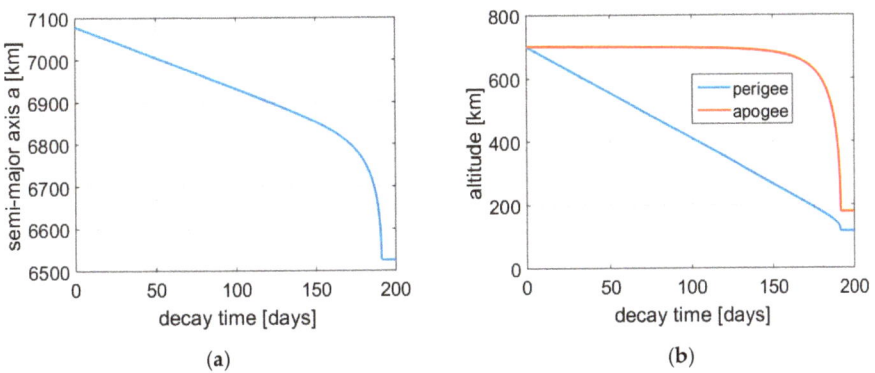

Figure 24. Orbital decay for an electric propulsion system that is activated only +/− 18° around the apogee: (**a**) semi-major axis; (**b**) apogee and perigee.

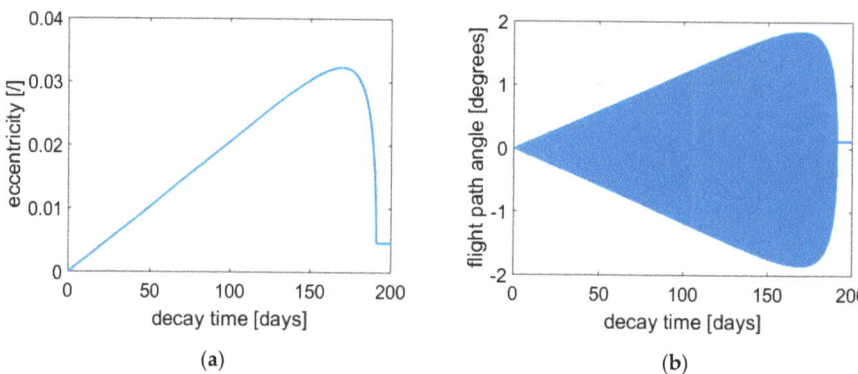

Figure 25. Orbital decay for an electric propulsion system that is activated only +/− 18° around the apogee: (**a**) eccentricity; (**b**) flight-path angle.

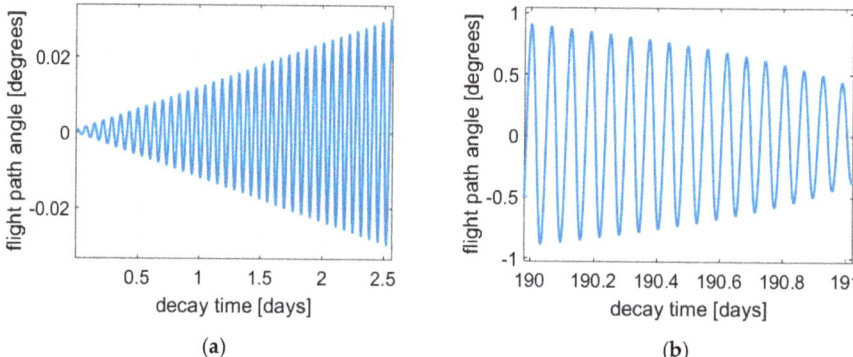

Figure 26. Flight-path angle during orbital decay for an electric propulsion system that is activated only +/− 18° around the apogee: (**a**) initial perigee lowering; (**b**) drag circularization.

In Figure 23, it is possible to see the propellant consumption with time. In Figure 23b, it is possible to see that the thruster is activated only around 10% of the time. In Figure 23a, it is possible to see that the propellant mass linearly increases with time (albeit at steps) and reaches the value of the half Hohmann transfer strategy. The little discrepancy appears only at low release altitudes as the atmospheric drag adds on the propulsion thrust helping in a further propellant mass saving. In Figure 24 it is possible to see that this partial activation of the electric thruster reduces the energy of the orbit (i.e., its semi-major axis) mainly through a decrease in the perigee altitude. Only when atmospheric drag comes into play does circularization occur. The eccentricity of the orbit increases (remaining very small) with the flight-path angle (Figure 25) until drag-induced circularization occurs. Figure 26 shows two details of the increase in flight-path angle (i.e., eccentricity) forced by the apogee thrusting phases (Figure 26a) and of the final circularization induced by the atmospheric drag (Figure 26b).

It is interesting to see how this strategy works for a fixed decay time. In this case, the shortest is the activation time and the highest should be the thrust in order to complete the decay in the same time. Apparently, the thrust should be inversely proportional to the activation time. This is correct as an order of magnitude estimate, but it is necessary to consider that as the activation time is decreased, the energy required to be drawn from the orbit is also cut by almost half.

Consequently, if the thrust is scaled as the inverse of the activation time (Figure 27a), the decay time follows the behavior of Figure 27b. The decay time for short activation times is almost half the one for a continuous operation.

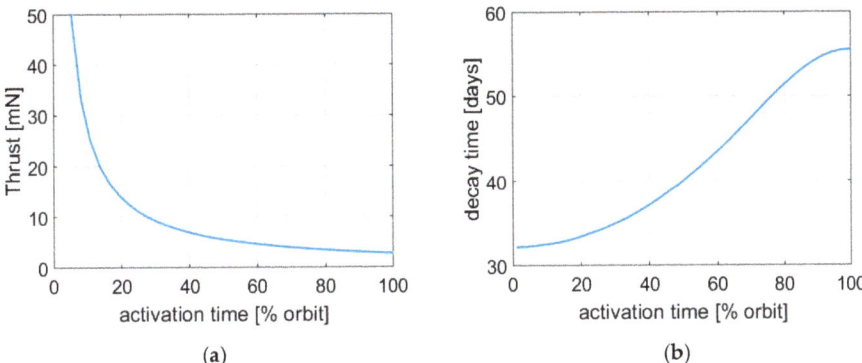

Figure 27. Deorbiting with an electric propulsion system that is activated for a fraction of the orbit: (a) thrust; (b) decay time.

As expected, the total thruster actuation time is equal to the total decay time for continuous thrusting and goes to zero as the activation time fraction goes to zero (Figure 28a). This mean that partializing the activation time could help being compliant with thruster firing time limitations.

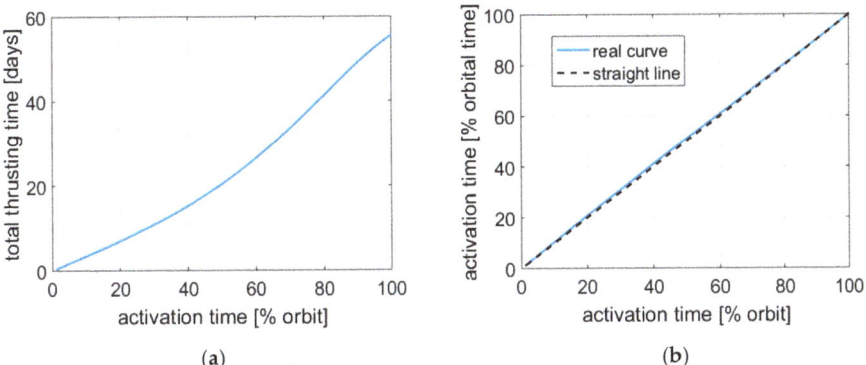

Figure 28. Deorbiting with an electric propulsion system that is activated for a fraction of the orbit: (a) total thrusting time; (b) activation time vs. orbital fraction.

As the eccentricity is very low there is negligible difference between true anomaly, eccentric anomaly and mean anomaly; the same stands for activation time in terms of fraction of orbit or fraction of orbital period (Figure 28b).

The mass of the thruster is proportional to the thrust of the system, and so increases as the activation time is reduced (Figure 29a). The propellant mass shifts from the one of a half Hohmann maneuver to the one of a full Hohmann maneuver if the transfer does not reach very low altitudes. In the latter case, drag comes into, play further reducing the propellant mass need (Figure 29b).

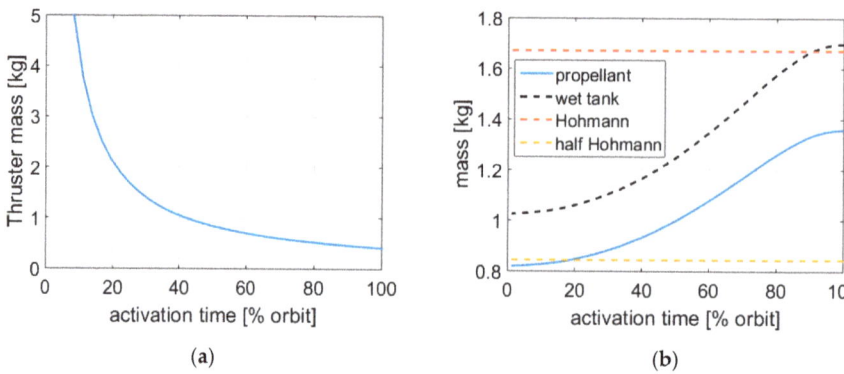

Figure 29. Deorbiting with an electric propulsion system that is activated for a fraction of the orbit: (a) thruster mass; (b) propellant-related mass.

The effect of drag is much more prominent for continuous thrusting as all the orbit is spiraling down at low altitudes so the high drag can act along all the orbit and not only near the perigee.

The total mass of the system is the sum of the thruster mass and the propellant-related mass. The first is decreases with the activation time, while the other does the opposite. Comparing Figure 29 it is clear that it makes little sense to narrow too much the thrusting time near the apogee (let us say below 20%) as the propellant savings are little while the mass of the thruster soars.

In Figure 30 the total mass of the propulsion system has been calculated adjusting the thrust exactly to obtain the same total decay time (so something slightly different than Figure 27). Different decay times from 1 month to 12 months have been considered. It is possible to see that the total mass reduces as the decay time is increased because the thruster mass is reduced. Moreover, an optimum point is present at a specific activation time that minimizes the total mass as a compromise between propellant mass and thruster mass. The optimum point shifts toward the left as the decay time is increased because of the lower impact of the thruster mass. A continuous thrusting strategy is not that far from the optimum mass, particularly for short total decay times. The results depend on the selected thruster parameters: $k = 0.25$, $Isp = 1000$ s and $\eta = 0.65$. A lower thruster efficiency or a higher Isp will shift the optimal point toward a continuous firing strategy.

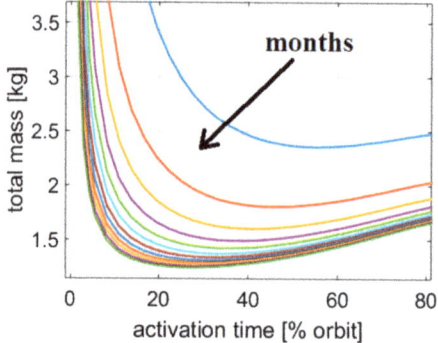

Figure 30. Deorbiting with an electric propulsion system that is activated for a fraction of the orbit. Total mass as a function of activation time, and parametric with total deorbit time (1–12 months).

4.5. Electrodynamic Tether

The tether mass is linearly dependent on the (inverse) of the decay time. The tether mass is also dependent on the orbital inclination (trough the sine, squared) and proportional to the total displacement (i.e., distance between the original and final orbit). Its behavior is very similar to that of an electric thruster, but no propellant is consumed, so the best solution is always to use it continuously during all the descent phase in order to minimize the required drag force and consequently tether size and mass. The behavior of the tether will be better illustrated in the following comparison subsection.

4.6. Comparison

It is now interesting to compare the different behavior and performance of the various deorbiting technologies. Figure 31 shows the time profile of the altitude for the different technologies sized in order to have the same decay time. The initial orbit is circular, the initial altitude is 680 km and the decay time is one year for all cases.

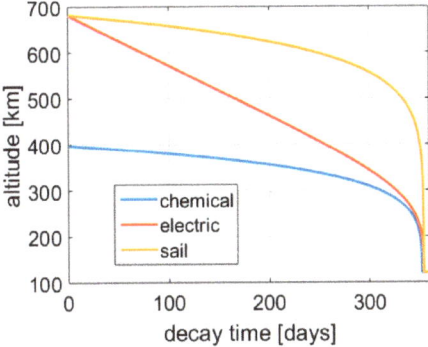

Figure 31. Comparison of different deorbit technologies' behavior, 680 km initial altitude. Orbital altitude with time.

The behavior of the electric thruster and the tether are the same if the thruster is fired continuously. Both systems induce a linear decrease in the orbital altitude until the drag prevails. The drag sail spends the majority of time near the initial altitude, and fall steeply near the end due to the atmospheric density profile with altitude. The chemical thruster immediately (compared with the total time) displaces the spacecraft to a new orbit at an altitude around 400 km to then fall slowly with the atmospheric drag in the same way the drag sail does at the higher altitude. This aspect is interesting to consider, as the decay altitude profile influences with time the way the system interacts with the other spacecrafts or debris. Even if the comparison is done for the same total decay time, there is a dramatic difference between spending almost one year in a crowded orbit or few hundred km below, for example.

A similar comparison can be also performed for an initial elliptical orbit. Again, comparing the deorbiting with a drag sail vs. a tether (or a continuously firing electric thruster), it is possible to recognize the different behavior. As the drag sail force is almost an exponential function of altitude, the decay is slower at the beginning and faster near the end. Moreover, the first phase is characterized by an orbit circularization, as the perigee (where the maximum force is applied) remains nearly constant while the apogee altitude decreases significantly. Near the end, the spacecraft spirals down. An increase of the drag sail area by a factor of ten causes a reduction of the deorbit time by the same amount (34 vs. 344 days in Figure 32a).

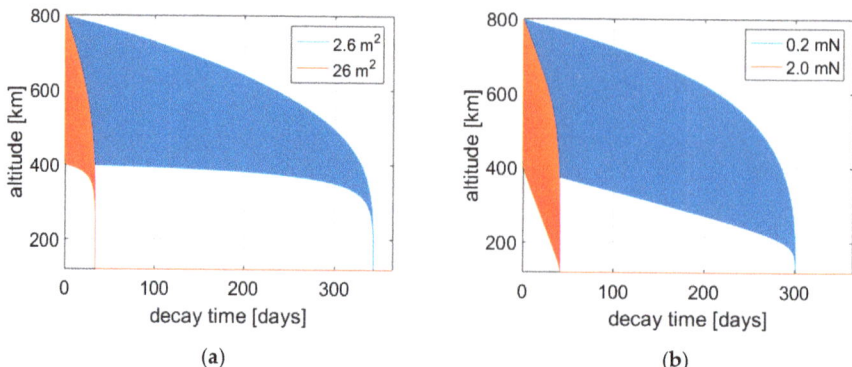

Figure 32. Deorbiting behavior with different technologies and a tenfold increase in the drag force (**a**) drag sail; (**b**) tether.

For the tether, the force is almost the same for the entire orbit and during the whole decay. The resulting effect is an almost parallel linear decrease in both the perigee and apogee altitudes. When the perigee starts to be low, the satellite drag adds to the tether force and the circularization process begin. In this phase the perigee fall is still linear while the apogee one is more exponential. Finally, the spiraling down occurs. The nonlinear sum of the satellite drag with the tether force is more relevant when the two forces are comparable, i.e., at low altitudes and low tether forces. For this reason, a tenfold decrease in the tether thrust from 2 mN to 0.2 mN induces a sublinear increase of the decay time from 42 to 301 days (Figure 32b).

Regarding the performance aspect, it is interesting to compare the deorbiting time at different altitudes for a similar mass budget. In the following Figures 33–35, a fixed mass budget of 4% has been imposed to the propellant-less systems, i.e., the sail and the tether, leaving the decay time to adjust consequently to the initial altitude. In the case of the electric thruster, the mass budget has been fixed equal to 4% mass for an initial altitude of 400 km (Figure 33). For higher altitudes the constant thruster mass has been kept the same and the propellant mass has been adapted to the higher delta-v, producing a slow variation (increase) of the total mass. The parameters for the electric propulsion are $\eta = 0.65$, $k = 0.12$ and $Isp = 3000$ s. For higher values of k and lower Isp the sensitivity of the total mass with the initial altitude increases. For the chemical propulsion system, the mass budget is mainly dependent on the propellant mass and thus the delta-v ($k = 0.12$, $Isp = 300$ s). As shown before, there is little room for the chemical system to trade mass for time. For this reason, the chemical propulsion mass is a strong function of the initial altitude.

All the technologies consider a full deorbit down to 150 km. In case of the chemical propulsion system, both n Hohmann transfer to 150 km and a half Hohmann transfer to 150 km are considered for simplicity. The two are not exactly equivalent as shown before, as the first will provide immediate deorbit while the second should require a slightly lower perigee to be exact, otherwise the deorbit time is a bit longer but still negligible compared with the other technologies (Figure 34b).

The delta-v of the Hohmann and half Hohmann are plotted in Figure 34a as a function of the altitude together with the delta-v required by the electric propulsion system. The electric propulsion system is considered spiraling down through a continuous firing. For the reasons just explained, the delta-v of the half Hohmann is around 50% the full Hohmann (while it will be more correct to stay on a 60% value for the same decay).

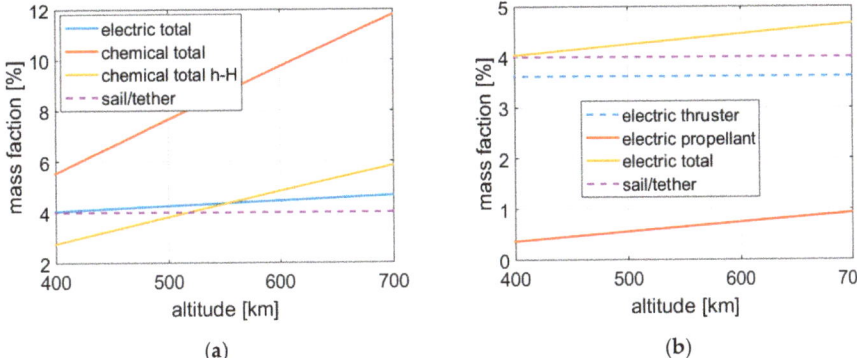

Figure 33. Deorbiting with different technologies from LEO up to 150 km altitude, mass budget: (**a**) full picture; (**b**) zoom-in.

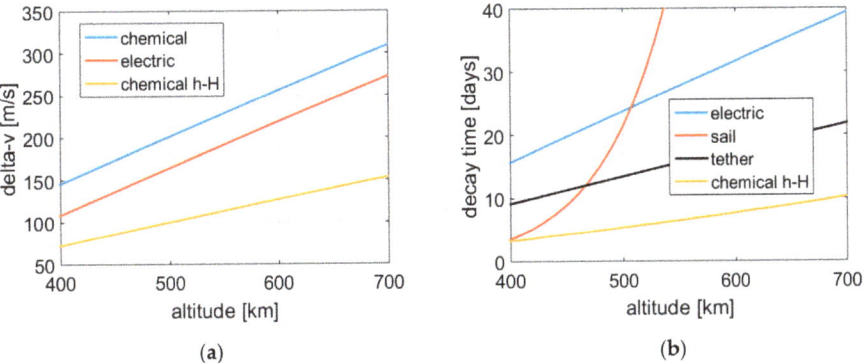

Figure 34. Deorbiting with different technologies from LEO up to 150 km altitude: (**a**) delta-v; (**b**) decay time. Tether data calculated for 45° orbital inclination.

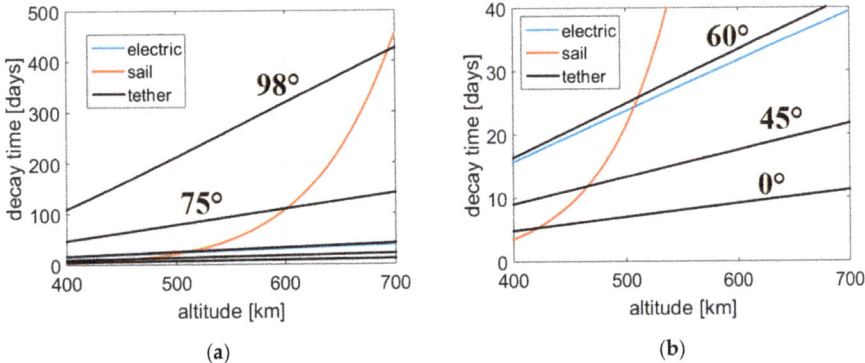

Figure 35. Decay time vs. altitude for different technologies: (**a**) full picture; (**b**) zoom-in. Tether data are parametric with orbital inclination.

The delta-v of the electric thruster is theoretically higher than the one of an impulsive full Hohmann transfer but, as stated earlier, at LEO altitudes the difference is negligible. However, as the descent with the electric thruster is slower, part of the delta-v is provided by the drag at the lowest altitudes (roughly the last 100 km, i.e., around 50 m/s), so the final result is the one plotted in Figure 34a and provides some propellant mass saving.

The decay times for the different technologies are shown in Figures 34b and 35. It is possible to see that the chemical propulsion system is always the fastest but also generally the heaviest, except for small displacements, particularly with the half Hohmann strategy.

The drag sail is very efficient at low altitudes but becomes dramatically inefficient at higher altitudes due to the exponential behavior of atmospheric density.

As already pointed out earlier, the electric propulsion system and the tether have a similar behavior, with a slow, almost linear variation of decay time with initial altitude for the same mass budget. The tether is theoretically the best system in the majority of cases, except for high orbital inclinations where it becomes much less efficient than the electric thruster.

A similar analysis has also been performed at higher altitudes. In this case, instead of a full deorbit, it has been considered to reposition the spacecraft to a final circular orbit with a 5-year natural decay time (slightly below 500 km). The drag sail has been excluded from the analysis because at this altitude the repositioning time becomes unsustainable.

The results (Figures 36 and 37) are analogous to the previous analysis. This time the delta-v of the full Hohmann transfer and the electric spiraling down are coincident, as the drag does not come into play at these altitudes (Figure 36a). For large displacements, the propellant mass budget of the chemical system grows to very high levels, and even the electric propulsion system is significantly affected.

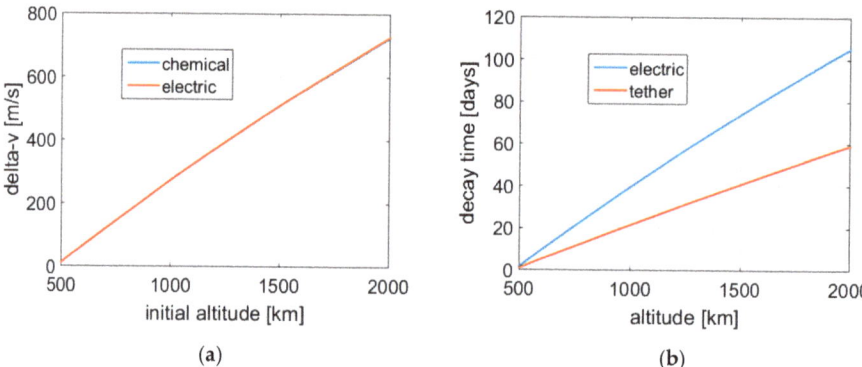

Figure 36. Deorbiting with different technologies from high altitudes in LEO up to the altitude of a 5-year natural re-entry: (**a**) delta-v; (**b**) active descent time. Tether data calculated for 45° orbital inclination.

The analysis has been repeated for a final natural decay time of 25 years. A 5-fold increase in the decay time provides only a few dozen km of higher final parking orbit altitude. For this reason, the results are almost equal to the previous ones (Figures 38 and 39).

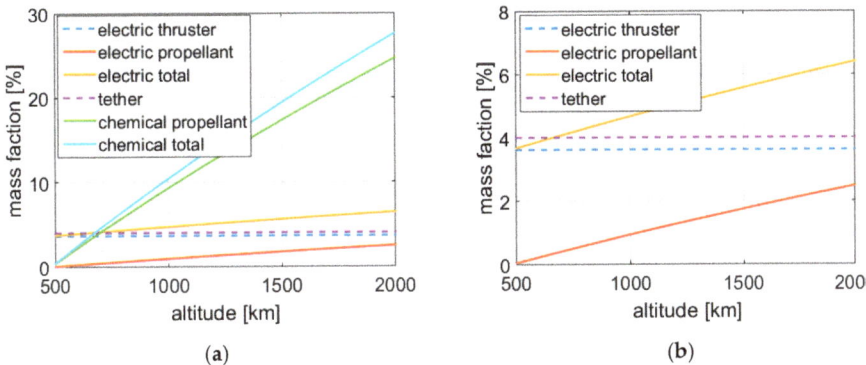

Figure 37. Deorbiting with different technologies from high altitudes in LEO up to the altitude of a 5-year natural re-entry, mass budget: (**a**) full picture; (**b**) zoom-in.

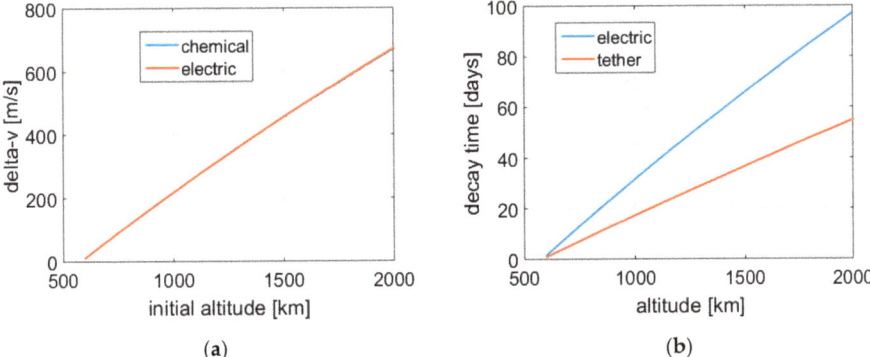

Figure 38. Deorbiting with different technologies from high altitudes in LEO up to the altitude of a 25-year natural re-entry: (**a**) delta-v; (**b**) active descent time. Tether data calculated for 45° orbital inclination.

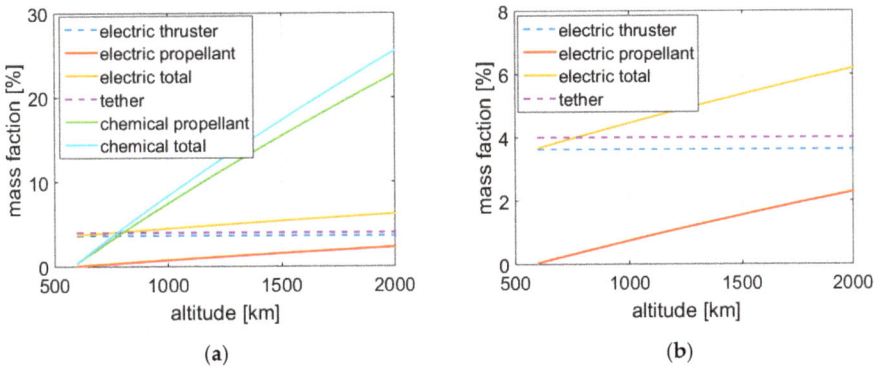

Figure 39. Deorbiting with different technologies from high altitudes in LEO up to the altitude of a 25 years natural re-entry, mass budget: (**a**) full picture; (**b**) zoom-in.

This shows how there is little relative loss to significantly improve the decay time of a spent satellite, unless it is are already near the parking orbit (where the relative loss soars but the absolute effort vanishes). This suggests the opportunity to impose a tighter limit on new launched spacecrafts, as the long-term benefits of doing so seem to justify the relatively small added effort.

4.7. Combinations

It has been shown that the chemical system is fast but heavier, while the drag sail is simple and efficient at low altitudes. It is interesting to combine these two technologies, using the chemical system to rapidly move a spacecraft out from a crowded region and displace it at a lower altitude where the drag sail can perform more efficiently.

As an example, the deorbit from an 800 km initial circular orbit has been considered. A chemical propulsion system performs a Hohmann transfer from 800 km to a certain lower altitude called release altitude. The corresponding mass budget is calculated. Then, from the final parking orbit, a drag sail is sized in order to complete the deorbit in a certain prescribed amount of time. Three cases have been considered: 3 months, 12 months/1 year (×4) and 4 years (×4 again), together with two techniques for the chemical system, a full Hohmann transfer to the release altitude and a half Hohmann transfer that lowers only the perigee to the release altitude. The results are presented in Figures 40–42.

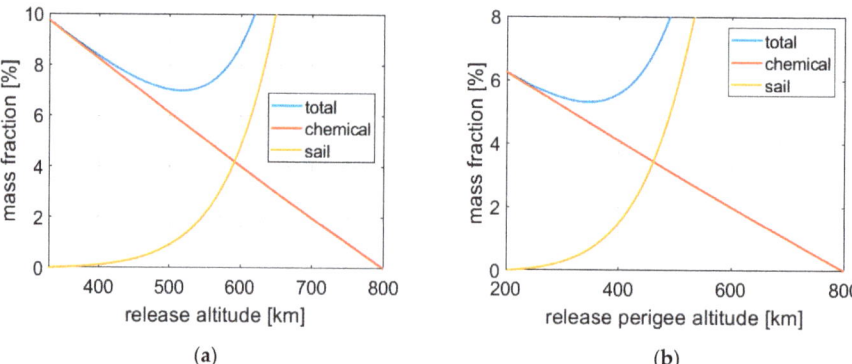

Figure 40. Deorbiting with the serial combination of a chemical propulsion system and a drag sail, mass fraction vs. release altitude, 3-month total duration: (**a**) full Hohmann; (**b**) half Hohmann.

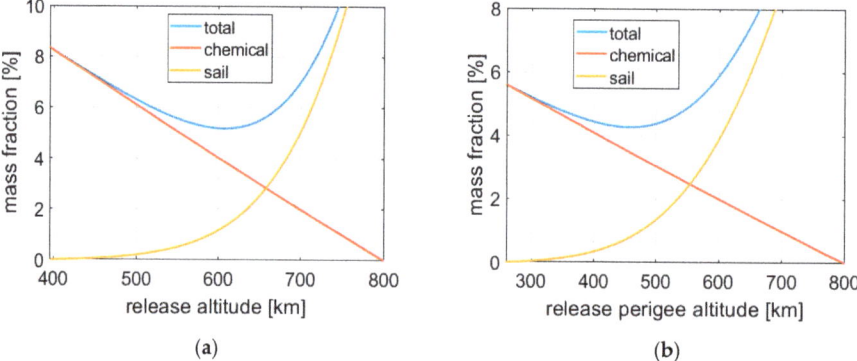

Figure 41. Deorbiting with the serial combination of a chemical propulsion system and a drag sail, mass fraction vs. release altitude, 1-year total duration: (**a**) full Hohmann; (**b**) half Hohmann.

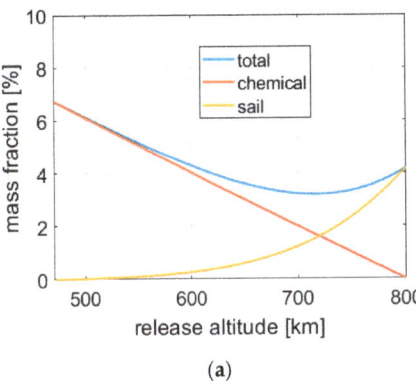 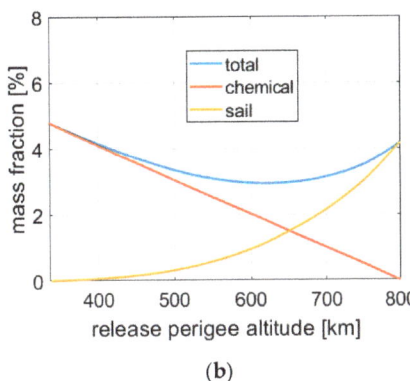

Figure 42. Deorbiting with the serial combination of a chemical propulsion system and a drag sail, mass fraction vs. release altitude, 4-year total duration: (**a**) full Hohmann; (**b**) half Hohmann.

In all cases, the chemical propulsion system mass is zero if the release altitude is the initial altitude (i.e., the deorbiting is performed only by the drag sail) and grows linearly as the release altitude is reduced. On the contrary, the mass of the drag sail is zero if the maneuver is completed only by the chemical system down to the corresponding natural decay altitude and increases exponentially as the release altitude is lifted up.

The mass of the chemical system is dependent only on the release altitude while the mass of the drag sail is dependent on the inverse of the total decay time as less area is required. The sum of the two masses has a minimum for a certain release altitude. The minimum mass altitude shifts to the right if the decay time is increased as the required sail mass is decreased. For the same reason, the benefits of a combined solution compared to a propulsive-only solution increase as the decay time is extended. On the contrary the benefits of a combined solution compared to a sail-only solution vanish as the decay time is extended.

Obviously, the half Hohmann technique guarantees lower masses and shift the optimum toward the chemical propulsion side (left). The combination of a drag sail with a chemical propulsion system is surely more complex than a single solution but becomes particularly interesting if a propulsion system is already on board and the drag sail operate in a full passive and efficient mode (regarding attitude) after deployment.

Figure 43 shows the altitude behavior with time of an optimal combined solution vs. the two single-system solutions. As already shown earlier, the initial displacement by a chemical thruster is favorable to rapidly moving the spacecraft on a less crowded orbit (unless the opposite occurs, as could be in some cases). Moreover, the huge mass savings of the half Hohman technique are balanced by the drawback of sweeping continuously a wide range of different altitudes during the decay.

Drag sails combined with tethers or electric thrusters are less attractive as the complexity starts to become relevant, and all three systems suffer from long decay times that will overlap in the mass budget.

A combination of a chemical system with a simple electrodynamic tether can be conceived in case a controlled re-entry is foreseen, but the total mass budget constraints do not allow for a full chemical solution.

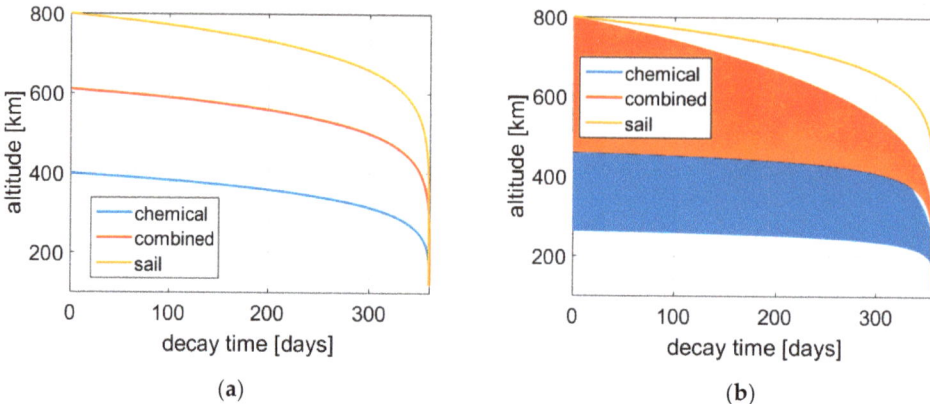

Figure 43. Deorbiting with the serial combination of a chemical propulsion system and a drag sail, altitude with time, 1-year total duration: (**a**) full Hohmann; (**b**) half Hohmann.

A combination of an electric thruster with a tether could be conceived of for the following reasons:

1. The tether in drag mode is able to generate power [40] that can be potentially used by the electric thruster;
2. The two systems can potentially be integrated, sharing some components like the cathode and the power management system, thus providing a synergistic effect;
3. A combined off-the-shelves system could be used at any orbital inclination with near-optimal performance. This is particularly interesting for constellations of satellites deployed in different orbital inclinations, where choosing a single combined commercial unit (instead of different ones, tether vs. electric, optimized for different inclinations) is advisable for mass production and integration.

4.8. Drag Compensation

A completely different approach to the deorbiting issue is to reverse the paradigm. Instead of taking care of deorbiting the spacecraft at the end of life, the satellite is placed directly at an altitude where the deorbiting is quick and assured by the natural aerodynamic decay. In this case, the propulsion system is not used to deorbit but to keep the satellite at the required altitude for its entire lifespan. This approach has important advantages. The first is the complete reliability of the deorbit process. While all the system previously discussed can fail preventing the proper deorbit of the spacecraft, in this case the deorbit is guaranteed. A failure of the propulsion system will lead to a failure of the mission and a premature re-entry of the spacecraft. From the standpoint of the space debris concern, this option is much more attractive.

It is interesting to compare the propellant mass required to keep a satellite at a certain altitude vs. the propellant mass required to deorbit the satellite. This is presented in the following Figures 44–50.

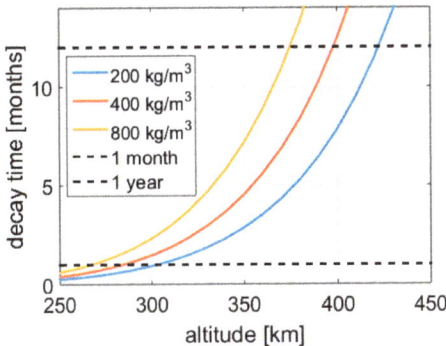

Figure 44. Natural decay time for a 50 kg spacecraft, parametric with satellite density.

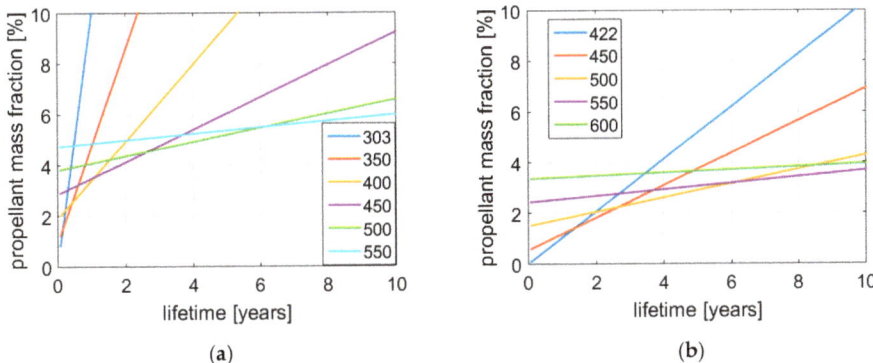

Figure 45. Propellant mass fraction vs. lifetime for station keeping and deorbiting (from flight altitude) with a chemical thruster (*Isp* = 300 s, satellite mass 50 kg, density 200 kg/m^3), parametric with flight altitude (km): (**a**) natural re-entry time of 1 month; (**b**) natural re-entry time of 1 year.

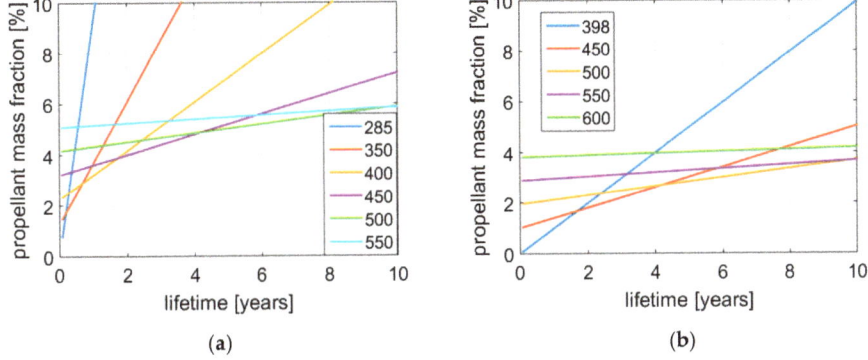

Figure 46. Propellant mass fraction vs. lifetime for station keeping and deorbiting (from flight altitude) with a chemical thruster (*Isp* = 300 s, satellite mass 50 kg, density 400 kg/m^3), parametric with flight altitude (km): (**a**) natural re-entry time of 1 month; (**b**) natural re-entry time of 1 year.

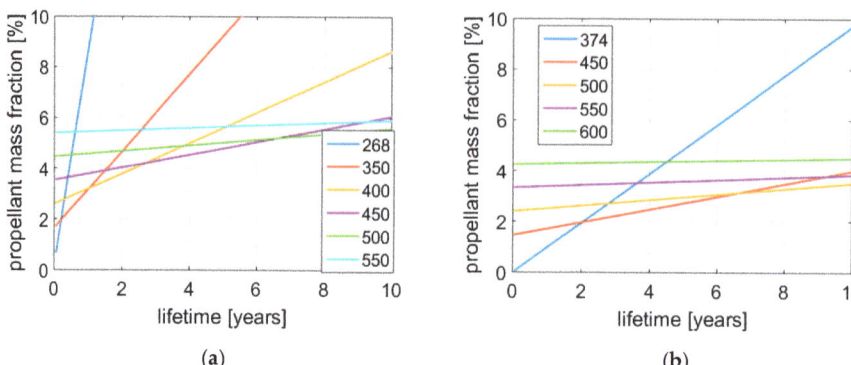

Figure 47. Propellant mass fraction vs. lifetime for station keeping and deorbiting (from flight altitude) with a chemical thruster (Isp = 300 s, satellite mass 50 kg, density 800 kg/m³), parametric with flight altitude (km): (**a**) natural re-entry time of 1 month; (**b**) natural re-entry time of 1 year.

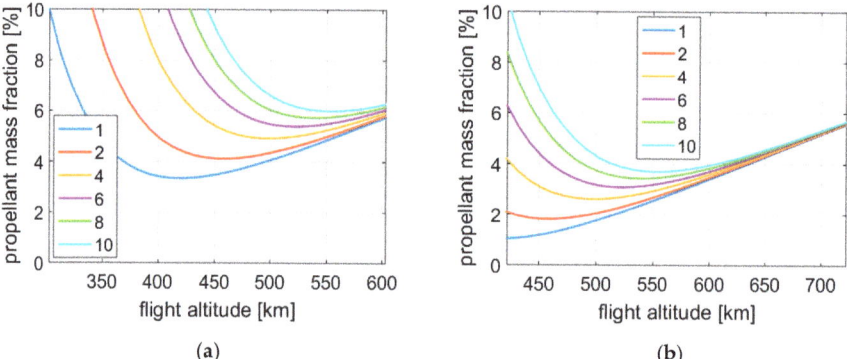

Figure 48. Propellant mass fraction for station keeping and deorbiting (from flight altitude) with a chemical thruster (Isp = 300 s, satellite mass 50 kg, density 200 kg/m³), parametric with satellite lifetime (years): (**a**) natural re-entry time of 1 month; (**b**) natural re-entry time of 1 year.

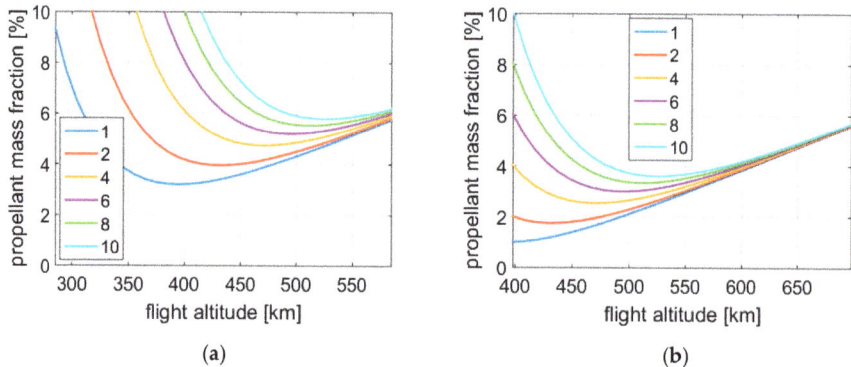

Figure 49. Propellant mass fraction for station keeping and deorbiting (from flight altitude) with a chemical thruster (Isp = 300 s, satellite mass 50 kg, density 400 kg/m³), parametric with satellite lifetime (years): (**a**) natural re-entry time of 1 month; (**b**) natural re-entry time of 1 year.

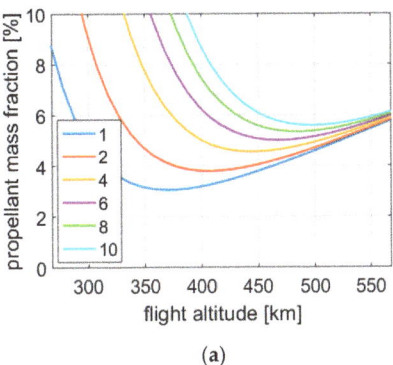 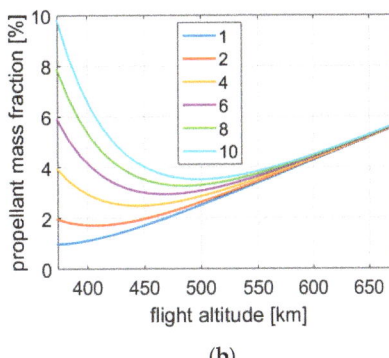

Figure 50. Propellant mass fraction for station keeping and deorbiting (from flight altitude) with a chemical thruster (Isp = 300 s, satellite mass 50 kg, density 800 kg/m^3), parametric with satellite lifetime (years): (**a**) natural re-entry time of 1 month; (**b**) natural re-entry time of 1 year.

Figure 44 shows the natural decay time of a spacecraft as a function of its initial altitude, parametric with the satellite density. The black lines correspond to the 1-month and 1-year limits. A higher ballistic coefficient allows a satellite to fly lower for the same decay time. From Figure 44, it is possible to infer the altitude required to naturally deorbit in 1 month (between 250 and 300 km) or 1 year (between 350 and 450 km).

The total propellant mass for the entire mission is the sum of the propellant mass for station keeping (i.e., drag compensation) at a certain altitude, plus the propellant mass required for repositioning at the altitude of the 1-month or 1-year limit. The propellant mass for drag compensation is linear with the satellite lifetime and (inversely) exponential with the flying altitude. Conversely, the propellant mass for deorbiting is independent from the satellite lifetime and almost linearly increases with the flying altitude. The resulting total propellant mass is a straight line that starts at a certain level (deorbiting needs) that is higher as the flying altitude increases. The slope of the straight line is related to the propellant mass consumed for station keeping. The lower the altitude, the higher the slope of the straight line because of the increased drag. The final pattern is visible in Figures 45–47. The difference between the left figures and the right figures is the final altitude, which corresponds to the 1-month and 1-year limit, respectively. The difference between the three figures (Figures 45–47) is the satellite density (i.e., ballistic coefficient).

The qualitative results are the same for all figures. Satellites flying at low altitude have a small deorbit need, but the propellant mass rapidly increases with satellite lifetime. On the contrary, spacecraft flying at higher altitudes have a significant deorbit need but the propellant mass varies little with the satellite lifetime as drag compensation requires little delta-v at those altitudes. Thus, short lifetimes favor low-flying satellites, while the opposite occurs for long lifetimes. Higher ballistic coefficients shift the optimal orbit to lower altitudes. Longer final decay times provide a higher minimum altitude and lower deorbit propellant mass needs.

Another way to present the results is shown in Figures 48–50. The same cases have been considered. This time the propellant mass fraction is presented as a function of the flying altitude, parametric with the satellite lifetime. It is possible to see that an optimal altitude exists that minimizes the total propellant consumption. Above the optimal altitude, increased deorbiting needs cause a higher propellant mass, while below, drag compensation is to be blamed.

The optimal altitude shifts to the right as the lifetime is extended because of the total impulse increase for drag compensation. The curves shift to the left as the ballistic coefficient is improved. As expected from the results highlighted in the previous figures, the results have very low sensitivity to the lifetime on the right of the graphs, where deorbiting need prevails, and a very high sensitivity on the left, where drag compensation prevails. Consequently, the optimal points for shorter lifetimes perform poorly for longer lifetimes, while the optimal points for longer lifetimes are not as bad for short lifetimes.

Finally, the effects of a different re-entry time limit (in this case 1 month vs. 1 year) are only to decrease the propellant mass requirements and set a higher minimum altitude threshold, but the curves' behavior seems the same with the same optimal altitudes.

Electric propulsion is particularly suited for drag compensation as the thrust (thus power) requirements are relatively low (without the thrust/time trade-off of deorbiting) and the high specific impulse limits the propellant mass burden [50].

It is important to highlight that flying at lower altitudes can have important side benefits other than the deorbiting aspects, particularly for Earth observation missions. In reality, these advantages are much more than simple side benefits, and they can already justify the choice of a lower-than-usual flying altitude by their own [51].

5. Other Topics

There are some other aspects that are interesting to be outlined in the framework of this paper, and they will be presented in this chapter.

5.1. Integrated vs. Independent System

The deorbiting device can be conceived of as an integrated or an independent sub-system of the spacecraft. In the first case, the satellite provides all the basic functions like attitude determination and control, communication, power and so on. This is the most common situation and provides generally the simplest, cheapest and lightweight solution. However, this option requires the satellite to stay alive and working up to the end of the mission, maybe not completely functional regarding the payload or other non-strictly necessary capabilities, but still capable of operating the deorbit correctly.

Approximately, the probability of the satellite to perform a successful re-entry is the product of the reliability of the deorbiting device/maneuver multiplied by the probability of the satellite to survive up to the end of the deorbit, which, as expected in general, in case the active part of the deorbiting is short compared to satellite lifetime, it is mainly related with the satellite surviving up to the decommissioning date. Some solutions can be considered inherently more reliable, for example a chemical propulsion system provides a short and thus less risky deorbit compared with a tether or an electric thruster, and a passive drag sail will also have a short active phase. For the reasons just highlighted, if the satellite has an average predicted lifetime with a certain uncertainty, it is necessary to end the mission and deorbit the satellite prematurely if a high success rate is sought, based on statistical analysis. The higher the lifetime uncertainty and the target success rate, the shorter will be the mandatory decommissioning time limit compared with the real average satellite lifetime.

One possible means to improve the effectiveness of this solution is to exploit the live monitoring and prediction of the satellite health. Thanks to modern artificial intelligence (AI) capabilities, it will be likely possible in the future to better assess the probability of failure of a specific spacecraft based on the story of its own health data compared with other similar assets. This could be particularly effective for large constellations of equal satellites.

In the opposite direction, an alternative solution is to equip the satellite with a fully autonomous deorbiting device that can be activated and operated successfully independently from the satellite, giving the possibility to wait until the satellite is dead. This could be particularly interesting for large systems with very long lifetimes that make huge profits for every (relatively small but absolutely massive) increase in operational time, and thus do not want to be prematurely decommissioned. This kind of system has been proposed, for example, by D-Orbit [52]. The issues of this solution are mainly two. The first drawback is the duplication of several satellite functions, which will bring added costs and weight. This is particularly true for small platforms, while probably much less of a deal for larger ones

(D-Orbit claims an added mass of few percent for a large GEO telecom sat). The second, very demanding challenge, is to fully guarantee the deorbit of a non-functional satellite in all (or almost all) the situations, particularly regarding damaged and or uncontrolled systems that have lost their attitude and are tumbling.

It is also possible that for some type of deorbiting devices a sort of intermediate solution is possible, where the main satellite can lose a significant but not complete part of its functions and still deorbit safely.

5.2. Active Debris Removal and Life Extension

While the bulk of the paper deals with the autonomous re-entry of a spacecraft and focuses on future objects launched into space, it is important to remember that there is a high chance that there will be the need to retrieve large objects already in space, particularly part of spent upper stages. To do that is necessary to perform an active debris removal (ADR) mission [36,37,53].

Active debris removal is by far more complex than the simple deorbiting of a satellite because the chasing spacecraft has to reach the target, approach it, grab it and to bring it back to the edge of the atmosphere or on a disposal orbit. The analysis of all these problematics is out of the scope of this paper and the reader is referred to the corresponding appropriate literature. What is worth mentioning here is that the propulsion/deorbiting systems should have several analogies with the one described previously, but with more demanding delta-v and functional requirements as the chasing spacecraft has to move first toward the target, doing proximity operations and deorbit it afterwards, sometimes capturing and deorbiting more than one object in the same mission. The technical advantage of an ADR platform is that the total lifetime of this system can be much shorter than that of a typical satellite, simplifying some engineering aspects. This is particularly true in case the target is a medium/large platform with a long lifetime.

ADR platforms could also highly benefit by strong synergies from future development in the so called in-orbit servicing market and in all the applications that require some sort of space tug. Another interesting related application is life extension, where the satellite lifetime is extended by the arrival of some kind of servicing platform.

It is also possible that future regulations will force ADR for satellites that have failed to deorbit. If the servicing market will achieve great success, it is even possible that the original satellites will not be equipped with a deorbiting device and then simply captured and disposed after death, booking the service (maybe in advance) from a commercial provider.

5.3. Type of Re-Entry

The re-entry of a spacecraft can be controlled or uncontrolled. In the latter, the system is deorbited without consideration on the actual re-entry corridor on the atmosphere. The uncontrolled re-entry simplifies the system design, and all the possible deorbiting devices, can be used, active or passive, fast or slow. On the contrary, for a controlled re-entry, it is necessary to have a system package that is able to direct the spacecraft toward a specific region of Earth.

Uncertainties in the orbit and attitude prediction are dominated by the large uncertainties in the atmospheric model, and in the solar and geomagnetic activity forecasts. For this reason, the final re-entry corridor can be predicted accurately only near the final orbit and thus controlled re-entry requires a quick and steep descent from a region of relatively thin density (where the satellite is easy to control) through the higher density layers of the atmosphere (where aerodynamic force and torque disturbances are dominant). Consequently controlled re-entry requires generally a chemical propulsion system, or a drag sail deployed at the right time at very low altitude (but not at high altitude, unless the area can be actively adjusted). Such a controlled re-entry under a relatively steep atmospheric incidence angle produces a confined ground impact area of break-up fragments which have survived the aerothermal heating. The ground impact area must be selected such that a tolerable residual risk to persons on ground can be achieved.

Controlled re-entry is mandatory for recoverable and/or reusable systems, which are rare at present but seem go be becoming more frequent in the next future, even for small platforms. In these cases, re-entry accuracy is even more important in order to allow safe landing and recover. Controlled re-entry is also necessary for expendable spacecrafts that produce particularly large and/or dangerous debris. NASA re-entry requirements [10] dictate the risk of human casualty anywhere on Earth due to a reentering debris with Kinetic Energy $KE \geq 15$ J be less than 1:10,000 (0.0001). This applies especially for denser materials, thicker structures, and more protected components.

An alternative possibility is design for demise [53–57]. In this case the expendable satellite is designed to burn completely in the atmosphere. Demisable designs will tend to favor lighter materials, thinner structures, and more exposed components. Design for demise allows the simplicity, reliability and multiple choice in deorbiting solutions of uncontrolled re-entry. However, design for demise requirements can partially conflict with design for survivability in space ones.

6. Conclusions

LEO satellite population is expected to grow dramatically in the coming years. To avoid a corresponding unacceptable increase in the rate of collision with debris or collision avoidance operations it is necessary to keep the LEO region clean. The current 25 year-limit, which corresponds to a natural decay from around 600 km, will become probably insufficient in the future and faster ways to deorbit are to be sought.

This paper focused the attention on the means for effectively deorbiting a spacecraft. Four technologies for deorbiting have been investigated: drag sail, chemical propulsion, electric propulsion, electrodynamic tether. Both simplified numerical models and full integration of the dynamic equations with a 4th order Runge–Kutta scheme have been used in order to compare the deorbit behavior of the different technologies and highlight their impact at system level.

Solar sails have been excluded from deep analysis as they have been shown to be inefficient for deorbiting, as the force they produce from about 4.5 $\mu N/m^2$ solar pressure is by far inferior to competing technologies, except maybe for drag sails at higher altitudes. In addition, they need to be controlled on each orbit for a very long time.

Drag sails are very effective at low altitude and long decay times, and they can be passive after deployment. Sail mass is almost inversely linear with decay time. However, they become inefficient and too slow at higher altitudes. With a relative mass between 1 and 4% of the satellite, they can deorbit a satellite in less than 5 years up to 700–800 km, and less than 25 years up to 800–1000 km.

Chemical propulsion allows for fast and controlled deorbit from any altitude but tends to be the heavier option if the time constraint is relaxed as this technology has limited sensitivity to deorbit time. Re-entry times of less than an hour can be achieved with a system mass fraction up to 20% at 1200 km.

Chemical propulsion can deorbit the satellite with a Hohmann transfer to a lower altitude or can lower its perigee with a half Hohmann maneuver. The second option allows for nearly 40% propellant savings but keeps the satellite orbiting in a wide range of altitudes unless the perigee is already sufficient for direct re-entry.

Electric propulsion can provide important mass savings compared with the chemical option, particularly for high delta-v (i.e., higher altitudes), thanks to its superior specific impulse. Delta-v losses for continuous firing have been shown to be negligible in LEO. However, the power (thrust) constraint forces a slower process, in the order of months. With mass fractions comparable to a drag sail, an electric propulsion system can provide much shorter deorbit times at higher altitudes, approximately starting from above 500 km altitude. For a fixed specific impulse, the electric propulsion system mass decreases asymptotically with the deorbit time down to the bare propellant tank wet mass.

The electric thruster can be fired continuously in order to provide a spiral decay of the orbit or operated only near the apogee to lower the perigee. The second option provides significant propellant mass savings at the expense of longer operating times or higher thruster size (i.e., mass and cost). Near-continuous firing guarantees the minimum total mass for short deorbit times (few months) at high specific impulses while lower duty cycles have a potential for up to 40% mass savings for long decay times (several months) at lower specific impulses. Duty cycles below 20% provide no benefits.

Electrodynamic tethers have the potential to be the most lightweight solution, especially for long decay time as their mass is inversely linear to decay time, with the exception of high orbital inclinations where they suffer from poor performance due to the alignment with the magnetic field. In fact, for the same mass budget, induced drag forces can be up to 2–3 times the one from a corresponding electric thruster for inclination up to 45°. Above around 60° parity is achieved, while near a polar orbit the thrust drops to near zero. They are also the less mature technology for the moment.

The most interesting combination is chemical propulsion with a drag sail, which are the most different between each other but both relatively simple. In fact, the first is fast and effective at any altitude (included higher ones) but becomes heavy for large displacements, while the second is slow but passive, lightweight and much more effective at low altitudes. A proper combination of the two technologies can provide a minimum mass solution that rapidly take out the spacecraft from the original (crowded) orbit and let it decay slowly from a lower altitude. For very short times the total mass approaches the one of a chemical-only system, while for very long times the total mass approaches the one of a sail-only system. For intermediate deorbit times on the order of 1 year savings around 30% are potentially possible.

The deorbit device can be integrated or independent. The first solution is generally the simplest, cheapest and lightweight but requires the satellite to stay alive up to disposal. An independent system is more challenging (and probably heavier and expensive in total) but could provide deorbit after satellite death, giving the possibility to exploit the full satellite lifetime. This is particularly interesting for large platforms with very long lifetimes. Health monitoring can help the integrated solution to adapt to the single satellite history.

Active debris removal is a possibly interesting future option both to retrieve current space junk but also for new spacecrafts that fail to deorbit or that do not carry a disposal system. This technology could go hand in hand with life extension solutions and the general new paradigm of commercial in-orbit servicing.

Re-entry can be controlled or uncontrolled. The first solution requires a high thrust solution, at least for the terminal part, and it is necessary for recoverable/reusable systems or spacecrafts that produce dangerous debris upon re-entry. The uncontrolled re-entry is a simpler and more flexible option but requires compliance with design for demise practices.

A completely different paradigm is drag compensation. In this casem no deorbit device is necessary and re-entry is guaranteed at the end of life as the satellite flies at very low altitudes. This solution requires a propulsion system for station keeping. Drag compensation is favored by shorter lifetimes (few years), higher specific impulses (e.g., electric propulsion) and high ballistic coefficients (i.e., slender design) and is particularly interesting for Earth observation missions where a lower altitude can improve the resolution by a factor of 2.

Plenty of options are available for deorbiting satellites in LEO, and this paper can be used as a starting point for mission/system design, trade-offs and preliminary selection.

The most important step forward currently seems to be the regulatory framework, imposing stricter rules and active mandatory sanctions.

Funding: This research received no external funding.

Institutional Review Board Statement: Not applicable.

Informed Consent Statement: Not applicable.

Data Availability Statement: Not applicable.

Acknowledgments: The author wants to thank Giulia Sarego (now at Technology for Propulsion and Innovation, T4i) for helping with the reference material. A thank you also to Daniele Pavarin (University of Padova and T4i CEO) for useful ideas.

Conflicts of Interest: The author declares no conflict of interest.

References

1. Space.com. Available online: https://www.space.com/spacex-starlink-satellite-collision-alerts-on-the-rise (accessed on 19 August 2022).
2. Parabolic Arc. Available online: http://parabolicarc.com/2022/07/17/swarm-dodges-collision-during-climb-to-escape-suns-wrath/#more-87593 (accessed on 19 August 2022).
3. Ars Technica. Available online: https://arstechnica.com/information-technology/2017/10/spacex-and-oneweb-broadband-satellites-raise-fears-about-space-debris/ (accessed on 19 August 2022).
4. Parabolic Arc. Available online: http://parabolicarc.com/2022/03/26/space-situational-assessment-2021-the-growing-menace-of-space-debris/ (accessed on 19 August 2022).
5. Kessler, D.J.; Cour-Palais, B.G. Collision frequency of artificial satellites: The creation of a debris belt. *J. Geophys. Res. Space Phys.* **1978**, *83*, 2637–2646. [CrossRef]
6. Karacalioglu, A.G.; Stupl, J. The Impact of New Trends in Satellite Launches on the Orbital Debris Environment. In Proceedings of the NASA: 8th IAASS Conference, Safety first, Safety for All, Melbourne, FL, USA, 18–20 May 2016.
7. ESA Space Debris Office. *ESA's Annual Space Environment Report*; GEN-DB-LOG-00288-OPS-SD; ESA: Darmstadt, Germany, 2020.
8. Wittig, M. Space Debris and De-Orbiting. MEW Aerospace. In *TU Symposium and Workshop on Small Satellite Regulation and Communication Systems*; Czech Republic: Prague, Czech, 2–4 March 2015.
9. SpaceX. Available online: https://www.spacex.com/rideshare/ (accessed on 19 August 2022).
10. ESA. Available online: https://www.esa.int/Enabling_Support/Operations/Space_debris_mitigation_the_case_for_a_code_of_conduct (accessed on 19 August 2022).
11. Hull, S.M. NASA Disposal Guidelines. 2013. Available online: https://ntrs.nasa.gov/citations/20130000278 (accessed on 19 August 2022).
12. IADC Steering Group and Working Group 4. IADC Space Debris Mitigation Guidelines. 2020. Available online: https://www.google.com/url?sa=t&rct=j&q=&esrc=s&source=web&cd=&ved=2ahUKEwjzwbyA7df6AhUL3KQKHQgHDDsQFnoECAwQAQ&url=https%3A%2F%2Fwww.iadc-home.org%2Fdocuments_public%2Ffile_down%2Fid%2F5249&usg=AOvVaw3Y9m5766xNnRyXuPdNPtW5 (accessed on 19 August 2022).
13. Parabolic Arc. Available online: http://parabolicarc.com/2022/02/24/spacexs-approach-to-space-sustainability-and-safety/ (accessed on 19 August 2022).
14. Parabolic Arc. Available online: http://www.parabolicarc.com/2022/07/20/neuraspace-creates-smart-traffic-management-solution-for-satellite-constellations/ (accessed on 19 August 2022).
15. Janovsky, R. End-Of-Life De-Orbiting Strategies for Satellites. In Proceedings of the 54th International Astronautical Congress of the International Astronautical Federation, the International Academy of Astronautics, and the International Institute of Space Law, Bremen, Germany, 29 September–3 October 2003. [CrossRef]
16. Cornara, S.; Beech, T.; Belló-Mora, M.; Martinez de Aragon, A. Satellite Constellation Launch, Deployment, Replacement, and End-Of-Life Strategies. SSC99-X-1. In Proceedings of the 13th Annual AIAA/USU Conference on Small Satellites, Logan, Utah, 23–26 August 1999.
17. Sanchez-Arriaga, G.; Sanmartin, J.R.; Lorenzini, E.C. Comparison of technologies for deorbiting spacecraft from low-earth-orbit at end of mission. *Acta Astronaut.* **2017**, *138*, 536–542. [CrossRef]
18. Steyn, W.H. *De-Orbiting Strategies*; Stellenbosch University: Stellenbosch, South Africa.
19. Mostafa, A.; El-Saftawy, M.I.; Abouelmagd, E.I.; López, M.A. Controlling the perturbations of solar radiation pressure on the Lorentz spacecraft. *Symmetry* **2020**, *12*, 1423. [CrossRef]
20. Abouelmagd, E.I.; Mortari, D.; Selim, H.H. Analytical study of periodic solutions on perturbed equatorial two-body problem. *Int. J. Bifurc. Chaos* **2015**, *25*, 1540040. [CrossRef]
21. Abouelmagd, E.I. Periodic solution of the two–body problem by KB averaging method within frame of the modified Newtonian potential. *J. Astronaut. Sci.* **2018**, *65*, 291–306. [CrossRef]
22. *U.S. Standard Atmosphere*; U.S. Government Printing Office: Washington, DC, USA, 1976.
23. Picone, J.M.; Hedin, A.E.; Drob, D.P.; Aikin, A.C. NRLMSISE-00 empirical model of the atmosphere: Statistical comparisons and scientific issues. *J. Geophys. Res. Space Phys.* **2002**, *107*, 1468. [CrossRef]
24. Parabolic Arc. Available online: http://parabolicarc.com/2022/06/14/space-flight-laboratory-announces-successful-deorbiting-of-nanosatellite-with-drag-sail-technology/ (accessed on 19 August 2022).
25. Black, A.; Spencer, D.A. DragSail systems for satellite deorbit and targeted reentry. *J. Space Saf. Eng.* **2020**, *7*, 397–403. [CrossRef]
26. Long, A.C.; Spencer, D.A. A Scalable Drag Sail for the Deorbit of Small Satellites. *J. Small Satell.* **2018**, *7*, 773–788.

27. Taylor, B.; Fellowes, S.; Dyer, B.; Viquerat, A.; Aglietti, G. A modular drag-deorbiting sail for large satellites in low Earth orbit. In Proceedings of the SciTech 2020, Orlando, FL, USA, 6–10 January 2020.
28. Sikes, J.D.; Ledbetter, W.; Sood, R.; Medina, K.; Turse, D. Keeping Low Earth Orbit Clean: Deorbit Analysis for an Articulating Boom Drag Sail. *Adv. Astronaut. Sci.* **2020**, *175*, 1–15.
29. Faber, D.; Overlack, A.; Welland, W.; van Vliet, L.; Wieling, W.; Tata Nardini, F. Nanosatellite deorbit motor. SSC13-I-9. In Proceedings of the 27th annual AIAA/USU Conference on Small Satellite, Logan, Utah, 12–15 August 2013.
30. Schonenborg, R.A.C.; Schöyer, H.F.R. Solid Propulsion De-Orbiting and Re-Orbiting. In Proceedings of the 5th European Conference on Space Debris, Darmstadt, Germany, 30 March–2 April 2009.
31. Schonenborg, R.A.C. Solid propellant de-orbiting for constellation satellites. In Proceedings of the 4th International Spacecraft Propulsion Conference, Chia Laguna, Italy, 2–9 June 2004.
32. Barato, F. Challenges of Ablatively Cooled Hybrid Rockets for Satellites or Upper Stages. *Aerospace* **2021**, *8*, 190. [CrossRef]
33. Barato, F.; Paccagnella, E.; Pavarin, D. Explicit Analytical Equations for Single Port Hybrid Rocket Combustion Chamber Sizing. *J. Propuls. Power* **2020**, *36*, 869–886. [CrossRef]
34. Barato, F.; Bellomo, N.; Pavarin, D. Integrated approach for hybrid rocket technology development. *Acta Astronaut.* **2016**, *128*, 257–261. [CrossRef]
35. Barato, F.; Grosse, M.; Bettella, A. Hybrid Rocket Residuals—An Overlooked Topic. AIAA 2014-3753. In Proceedings of the 50th AIAA/ASME/SAE/ASEE Joint Propulsion Conference & Exhibit, Cleveland, OH, USA, 28–30 July 2014. [CrossRef]
36. Tadini, P. Hybrid Rocket Propulsion for Active Removal of Large Abandoned Objects. Ph.D. Thesis, Politecnico di Milano, Milan, Italy, 2014.
37. Tonetti, S.; Cornara, S.; Faenza, M.; Verberne, O.; Langener, T. Feasibility Study of Active Debris Removal Using Hybrid Propulsion Solutions. In Proceedings of the Stardust Global Virtual Workshop II, Southampton, UK, 19–22 January 2016.
38. *MIT OpenCourseWare 16.522 Space Propulsion*; Springer: Cambridge, MA, USA, 2015; Available online: http://ocw.mit.edu (accessed on 19 August 2022).
39. Sarego, G.; Olivieri, L.; Valmorbida, A.; Bettanini, C.; Colombatti, G.; Pertile, M.; Lorenzini, E.C. Deorbiting Performance of Electrodynamic Tethers to Mitigate Space Debris. World Academy of Science, Engineering and Technology. *Int. J. Aerosp. Mech. Eng.* **2021**, *15*, 185–191.
40. Bilen, S.; McTernan, J.; Gilchrist, B.; Bell, I.; Voronka, N.; Hoyt, R. Electrodynamic Tethers for Energy Harvesting and Propulsion on Space Platforms. AIAA 2010-8844. In Proceedings of the AIAA SPACE 2010 Conference & Exposition, Anaheim, CA, USA, 30 August–2 September 2010.
41. Olivieri, L.; Mantellato, R.; Francesconi, A. A Tethered Space Tug Concept Demonstration for Active Debris Removal Missions. In Proceedings of the ESA GNC 2017, Salzburg, Austria, 29 May–June 2017.
42. Thethers Unlimited CubeSat Terminator Tape Brochure. Available online: https://www.tethers.com/ (accessed on 19 August 2022).
43. Thethers Unlimited NanoSat Terminator Tape Brochure. Available online: https://www.tethers.com/ (accessed on 19 August 2022).
44. Aurora Propulsion Technologies APB-S Brochure. Available online: https://www.aurorapt.fi (accessed on 19 August 2022).
45. Montenbruck, O.; Gill, E. *Satellite Orbits: Models, Methods and Applications*; Springer: Berlin, Heidelberg, 2001. [CrossRef]
46. Chobotov, V.A. *Orbital Mechanics*, 3rd ed.; AIAA Education Series; American Institute of Aeronautics and Astronautics, Inc.: Reston, VA, USA, 2008.
47. Vallado, D.A. *Fundamental of Astrodynamics and Applications*, 2nd ed.; Microcosm Press: El Segundo, CA, USA; Kluwer Academic Publishers: Boston, MA, USA, 2001.
48. GOCE ESA's Gravity Mission. Available online: https://www.esa.int/Applications/Observing_the_Earth/FutureEO/GOCE (accessed on 19 August 2022).
49. Klinkrad, H.; Sanchez-Ortiz, N. Collision Avoidance for Operational ESA Satellites. ESA SP-587. In Proceedings of the 4th European Conference on Space Debris, Darmstadt, Germany, 18–20 April 2005.
50. Barato, F.; Trezzolani, F.; Manente, M.; Pavarin, D.; Andreussi, T.; Andrenucci, M. Electric Propulsion Technology for Low Earth Mission of Micro/Nano-Satellites. In Proceedings of the International Astronautic Conference 2015, Jerusalem, Israel, 12–16 October 2015.
51. Bertolucci, G.; Barato, F.; Toson, E.; Pavarin, D. Impact of propulsion system characteristics on the potential for cost reduction of earth observation missions at very low altitudes. *Acta Astronaut.* **2020**, *176*, 173–191. [CrossRef]
52. Rossettini, L. Technical Day on "Deorbiting Strategies". Available online: https://www.dorbit.space/ (accessed on 1 December 2015).
53. Innocenti, L. *Clean Space*; ESA Clean Space Office: Darmstadt, Germany, 2019.
54. Astromaterials Research & Exploration Science NASA Orbital Debris Program Office. Available online: https://orbitaldebris.jsc.nasa.gov/reentry/ (accessed on 19 August 2022).
55. Eggen, N.; Tiago Soares, T.; Innocenti, L. Containment Methods for The Atmospheric Reentry of Satellites. In Proceedings of the First Int'l. Orbital Debris Conference, Sugar Land, TX, USA, 9–12 December 2019.

56. Trisolini, M.; Lewis, H.G.; Colombo, C. On the Demisability and Survivability of Modern Spacecraft. In Proceedings of the 7th European Conference on Space Debris, Darmstadt, Germany, 18–21 April 2017; Flohrer, T., Schmitz, F., Eds.; The ESA Space Debris Office: Darmstadt, Germany, 2017. Available online: http://spacedebris2017.sdo.esoc.esa.int (accessed on 1 June 2017).
57. Waswa, M.B.P. Spacecraft Design-for-Demise Strategy, Analysis and Impact on Low Earth Orbit Space Missions. Master's Thesis, Massachusetts Institute of Technology, Cambridge, MA, USA, 2009.

Article

De-Orbit Maneuver Demonstration Results of Micro-Satellite ALE-1 with a Separable Drag Sail

Kohei Takeda [1], Toshinori Kuwahara [1,*], Takumi Saito [1], Shinya Fujita [1], Yoshihiko Shibuya [2], Hiromune Ishii [2], Lena Okajima [2] and Tetsuya Kaneko [3]

[1] Department of Aerospace Engineering, Tohoku University, Sendai 980-8579, Japan
[2] ALE Co., Ltd., Tokyo 105-0012, Japan
[3] Nakashimada Engineering Works, Ltd., Hirokawa 834-0196, Japan
* Correspondence: toshinori.kuwahara.b3@tohoku.ac.jp

Abstract: ALE-1, a micro-satellite created for the demonstration of artificial shooting stars, required orbital descent before mission execution due to safety aspects in orbit. ALE-1 utilized a drag sail called SDOM (Separable De-Orbit Mechanism) for a passive de-orbit maneuver, which was successfully completed, lowering the orbit from about 500 km down to about 400 km. This paper summarizes the detailed history of satellite operation and the results of the de-orbit maneuver demonstration during the past three years. Although the SDOM sail faced difficulty in keeping the desired deployed shape of the drag sail due to mechanical troubles, by letting the sail be a drag flag instead, it could still deliver a meaningful de-orbit performance to allow the satellite to successfully lower the orbit as planned. The de-orbit effect of the drag flag was evaluated using comparisons between orbit propagation simulations and the actual orbit transition flight data provided in the form of TLE (Two-Line Element) sets. Through this study, it is demonstrated that the SDOM can provide orbit transfer capabilities for satellites. Furthermore, the de-orbit performance of the drag flag can be evaluated, which could be an important reference for the future implementation of de-orbit devices to solve space debris problems.

Keywords: micro-satellite; drag sail; de-orbit; time-of-flight camera

1. Introduction

ALE-1 is a micro-satellite jointly developed by Tohoku University and ALE Co., Ltd. (Tokyo, Japan), and it was launched by Japanese Epsilon Launch Vehicle No. 4 on 18 January 2019 as part of the first implementation of JAXA's Innovative Satellite Technology Demonstration Program [1,2]. The mission of ALE-1 is to demonstrate the technology of artificial shooting star generation, and it is still operational in orbit. ALE-1 required a descent of about 100 km from the initial orbit altitude of about 500 km due to safety aspects in orbit, which was successfully accomplished with a passive drag sail de-orbit device called Separable De-Orbit Mechanism (SDOM). SDOM is a passive de-orbit system that uses atmospheric drag in the low Earth orbit, and it was jointly developed by Nakashimada Engineering Works Ltd. and Tohoku University. Although it takes a long time to de-orbit, it is a simple thin-film deployment system that is actuated by mechanical strain energy, and it is expected to be a solution to the space debris problem, which has been a significant concern in recent years [3,4].

We have reported on the development and operational progress of SDOM and ALE-1 in various ways. In particular, in 2022, we published a paper on the status of system trouble that occurred in SDOM and its impact on the orbit descent [5]. Two of the four corners of the square-shaped drag sail of SDOM were unexpectedly loosened, resulting in a drag flag. This system trouble seemed to decrease SDOM's performance as a de-orbit mechanism, and we discussed the prediction of a possible significant delay in the orbit descent.

Despite the trouble, however, SDOM was subsequently able to fulfill its role in orbit descent and revealed that the effect of the trouble was minimal. Moreover, SDOM was successfully separated from the ALE-1 satellite at the end of 2022 as intended, allowing the micro-satellite to stay in the target orbit for a longer time period. In this paper, we review the history and results of the orbital operation of ALE-1 in terms of various aspects and evaluate the descent performance of the SDOM based on the comparisons between orbit propagation simulations and the actual orbit transition record provided in the form of TLE (Two-Line Element) history.

2. Mission and System

2.1. ALE-1

Figure 1 shows the appearance of the flight model of the micro-satellite ALE-1 together with the positions of the related components, and Table 1 provides its specifications. The main mission of ALE-1 is to demonstrate the technology of an "artificial shooting star" generation by ejecting small metal pellets from orbit to Earth, where interaction with the atmosphere will cause a luminous phenomenon. This technological demonstration has two aspects: entertainment and scientific observation. The entertainment aspect involves the development of a service that allows people to enjoy shooting stars from any ground position at any given time. The scientific observation aspect aims to gather data on luminous phenomena in the upper atmosphere, which can be used to investigate the characteristics of the upper atmosphere and the re-entry behavior of small objects [6–9].

Table 1. Specifications of ALE-1.

	Parameters	Values
Satellite	Mass [kg]	68.16
	Dimensions [mm]	440 × 500 × 539
SDOM	Mass [kg]	3.88
	Dimensions [mm] (Stored configuration)	277 × 211 × 222
	Film size [mm]	2500 × 2500
	Thickness of the film [μm]	25

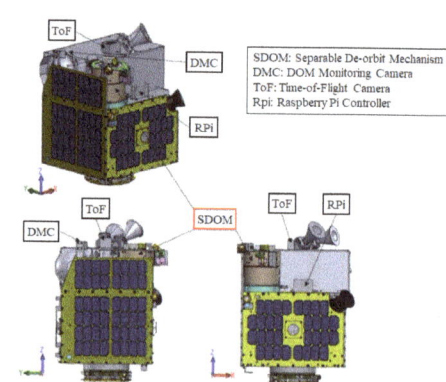

Figure 1. Micro-satellite ALE-1 and its component configuration.

In order to prevent collisions with other spacecraft in orbit, it is crucial to ensure that the pellets are carefully ejected. This is especially important for the International Space Station (ISS), which operates at an altitude of 400 km and houses critical experimental facilities and astronauts. To minimize the risk of any pellet impact with the ISS, ALE-1 needed to transfer from its initial orbit at an altitude of 500 km to a lower altitude below 400 km prior to commencing its main mission [10–12].

As the orbit transfer device, we installed an SDOM, which will be described in detail in the next section. The SDOM utilized a sail to increase the atmospheric drag to lower the orbit altitude and separated the sail section after ALE-1 had descended about 100 km, allowing it to remain in the lowered orbit for an extended period.

2.2. SDOM: Separable De-Orbit Mechanism

SDOM is a passive de-orbit system that was specifically developed for ALE-1. This system uses an aluminized polyimide thin film as an atmospheric drag sail and represents a new model of the de-orbit mechanism (DOM) that Nakashimada Engineering Works, Ltd. and Tohoku University have been collaboratively developing since 2010 [13,14]. The appearance of the deployed DOM is illustrated in Figure 2. The SDOM system drags down satellites using the atmospheric drag that acts on the thin film, allowing satellites to re-enter the Earth's atmosphere. Several models of DOM have been already operated in orbit, providing us with important engineering findings [15,16].

SDOM consists of the following components:

- A cylindrical container to ensure safety during rocket launch;
- A thin film deployment mechanism that acts as a drag sail to lower the orbit;
- A boom to keep the thin film away from the satellite so that it does not block the antenna or solar panels;
- A system to separate the drag sail from the container after reaching a predetermined target altitude.

Figure 2. Deployed configuration of DOM: De-Orbit Mechanism.

SDOM is designed to have five operational phases to ensure its proper functioning, as depicted in Figure 3 [5]. Phase 0 represents the safe storage of all components within the cylindrical body without deployment. In phase 1, the lid is opened. In phase 2, the boom is extended to position the drag sail mechanism 2.5 m away from the satellite's main structure. In phase 3, a 2.5 m × 2.5 m thin film is deployed to initiate de-orbit. Upon reaching a predetermined target altitude (in the case of ALE-1, less than 400 km), the film and boom are separated on command from the ground station to prolong the satellite's orbital lifetime at that altitude (phase 4). The cylindrical container remains attached to the satellite body.

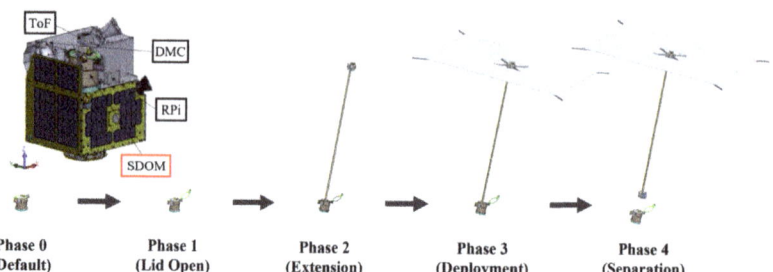

Figure 3. Operational phases of SDOM.

3. History of Operation

3.1. Mission History

In this section, the operational history of ALE-1 and SDOM is discussed using images from the TOF (Time-of-Flight) camera onboard ALE-1. The mission history is shown in Figure 4, and the corresponding TLE history of ALE-1 is shown in Figure 5.

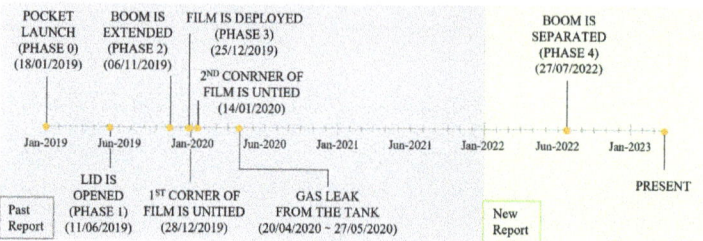

Figure 4. Mission history of ALE-1 and SDOM.

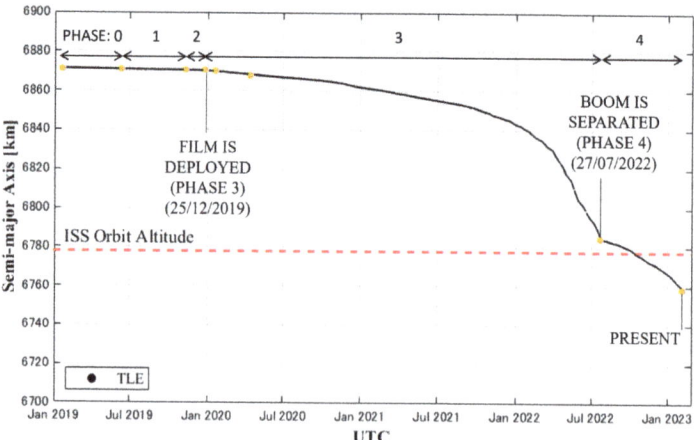

Figure 5. Orbital altitude history of ALE-1 based on TLE: Two-Line Element. The semi-major axis of the orbit is plotted together with the mission history.

The separated SDOM sail is unlikely to generate additional space debris, as its area-to-mass ratio is large enough to de-orbit itself for a very short orbital period, ultimately being burned up during the re-entry. One reference for this technology is the 1U-sized CubeSat FREEDOM, which was operated in 2017 by Nakashimada Engineering Works, Ltd. and Tohoku University [17]. FREEDOM carried a DOM with a film size of 1.5 m × 1.5 m. Based on the orbit history of FREEDOM, it took about 22 days to re-enter the Earth's atmosphere from an altitude of 400 km [13]. According to this background, the separated part of the SDOM is expected to re-enter the atmosphere within a few days.

A detailed description of the mechanisms and satellite subsystems of ALE-1 can be found in [7,18,19]. Each phase transition is initiated by stored commands that are uplinked from the ground station after the required status checks for each phase transition have been verified [11].

After ALE-1 entered orbit on 18 January 2019, the satellite was operated and underwent orbital functional verification. The SDOM mission started on 11 June 2019 when its lid was opened. Shortly thereafter, the boom extension was initiated, but it could not be confirmed as intended. After 5 months of regular observations, the boom extension was finally confirmed on 6 November 2019, which initiated phase 2. The deployment of the thin film was postponed for an additional 50 days to allow for the detailed evaluation of the gravity gradient effects acting on the satellite system. The film was then deployed on 25 December 2019 [5].

Based on the initial estimation, ALE-1 with the SDOM deployed was expected to descend to the ISS orbital altitude in about 650 days [18]; in contrast with the estimation, however, the deployed SDOM experienced a series of problems. On 28 December 2019, one of the film's four corner connections was found to be damaged, which left only half the area of the film effective for drag. Approximately two weeks later, another corner connection was lost, effectively changing the film from a drag sail to a drag flag. This issue is discussed in more detail in the next section. In addition, between 20 April 2020 and 27 May 2020, a gas leak occurred from a tank installed for the artificial shooting star mission; the gas leak was caused by a malfunction of the gas output control system, and the malfunction was resolved when the control system was restarted. After that, operations went smoothly, and on 27 July 2022, when the orbital altitude was confirmed to be below 400 km, the sail section of SDOM was separated from ALE-1. The SDOM separation was observed by the DMC (DOM Monitoring Camera) as illustrated in Figure 6. ALE-1 completed its descent, albeit about a year later than predicted, and is still flying today, maintaining its orbital altitude.

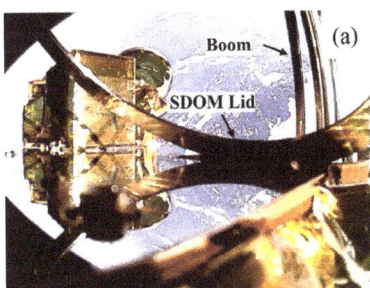 

Figure 6. The separation of the boom and film of SDOM, as observed by DMC. (**a**) Before SDOM is separated. (**b**) After SDOM is separated.

3.2. Trouble of SDOM

The SDOM system broke two of the film's four corner connections in late 2019 and early 2020, effectively changing the film from a drag sail to a drag flag; the exact reason for the broken connections in these corners is still unknown. However, based on the observations, it is more likely that the Dyneema wires in the connections were untangled rather than severed. This situation could be observed by the DMC, as illustrated in Figure 7. The DMC was originally installed to monitor the conditions of the SDOM lid and boom extension.

ALE-1 is also equipped with a camera, called the Time-of-Flight (TOF) Camera System, for observing the SDOM film. This camera system collects distance information to the sail surface and downlinks the data to the ground to allow analyses to obtain better insight into the three-dimensional dynamic behavior of the boom and drag sail in space [20]. In fact, we reported the results of our estimation of the shape of the SDOM when deployed in space in 2022 [21].

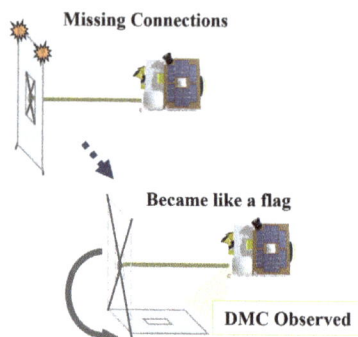

Figure 7. Disconnected SDOM film and its relative attitude as observed by DMC.

Now, the SDOM film with the two missing connections exhibited a characteristic behavior that was observed by the onboard cameras. It was determined that the film was exhibiting small movements around an axis through the remaining connections. Most of the time, the film remained on the opposite side of the DOM deployment plane in relation to the satellite, but there were instances where it was observed to move to the front side of the deployment plane. Figure 7 captured the film at the moment it unintentionally entered the field of view of the DMC while being positioned on the front side of the DOM deployment plane. The photograph clearly shows that one corner of the film is disconnected; the subsequent DMC photograph did not show the film at all. From this investigation, it was determined that the film can freely move around depending on the relative motion between the satellite's main structure.

Although the SDOM unintentionally became a drag flag, the orbital history shows that it was nevertheless effective for the orbital descent. This suggests that the SDOM, as it trailed like a cloak, experienced a certain level of atmospheric drag, which was expected to reduce the velocity of ALE-1. In this study, we examine the impact of this drag flag on the orbital descent by comparing the descent simulations and orbital history.

4. Investigation on Orbital Decent

4.1. Parameters of Orbit Analysis

We performed numerical simulations on the trajectory of ALE-1 with the SDOM film deployment. In general, the perturbed acceleration of a flying object in the upper atmosphere can be described by the following equation:

$$a_{\text{drag}} = -\frac{1}{2} C_D \rho \frac{A}{m} v_{\text{rel}}^2 \qquad (1)$$

where ρ is the atmospheric density, C_D is the drag coefficient, A is the cross-sectional area of the satellite, m is the mass of the satellite, and v_{rel} is the relative velocity against the atmosphere. The mass of ALE-1 is 68.16 kg (65.29 kg after SDOM separation), as shown in Table 1, which is measured before launch. The C_D is a dimensionless quantity; for a satellite, it is commonly set to be 2.2 [22]. Although A is 6.25 m² at maximum when SDOM is deployed, the cross-sectional area value cannot be measured in the situation because the film is in a drag flag state. In addition, because C_D varies with the shape of the drag flag as well as the angles of attack, it is difficult to make a reliable assumption on the value of C_D. It is also difficult to calculate the C_D due to the unknown shape of the drag flag for the long period of the orbit transfer. Thus, C_D and A need to be handled as unknown parameters, which act as the limitation of the analysis.

According to this background, we decided to use the combination of these parameters $C_D \times A$ as an evaluation index. The range of $C_D A$ for the simulation analysis is determined based on the traditional fixed value of $C_D = 2.2$ and the mean cross-sectional area of ALE-1. The mean cross-sectional area of a tumbling satellite can be approximated as the sum of the projected areas in the six orthogonal directions (plus and minus directions in each of the three orthogonal axes) divided by four. As this value is about 3.4 m² for ALE-1 with the fully deployed SDOM and about 0.4 m² without SDOM, we set $C_D A_{ave}$ to be in the approximate range of 0.80–7.43.

For the initial orbit conditions, the TLE of ALE-1 on 28 May 2020 was used. The reason for not using the TLE immediately after the SDOM deployment is to avoid the influence of the gas leak, which caused an instantaneous increase in orbit altitude. Orbit propagation was performed by using the fourth-order RungeKutta method, with the solar and lunar mass perturbations, atmospheric drag, solar radiation pressure, and Earth gravity field being based on a non-spherical central body model [23]. The environmental models applied to the simulations are listed in Table 2, and the parameters used are summarized in Table 3. The atmospheric density model was calculated based on the parameters available on 1 February 2023. The simulator used for the analysis is the MEVIμS system, which was developed by our research team [24,25].

Table 2. Environmental models of the orbit analysis.

Environment	Model	References
Earth gravity model	EGM-08 (degree, order) = (40,40)	Ref. [22]
Sun and Moon model	DE431	Ref. [22,23]
Atmosphere model	NRLMSISE-00	Ref. [26]

Table 3. Parameters of the orbit analysis.

Items	Parameters	Value	References
Atmospheric drag	C_D	-	Ref. [22]
	A [m²]	-	-
	$F_{10.7}$	Variable	Ref. [26]
	$F_{10.7\ 81\ days}$	Variable	Ref. [26]
Solar radiation	c_R [-] (reflectivity)	1.0	Ref. [22]
	P_{srp} [N/m²] (solar-radiation pressure)	4.54 E-6	Ref. [22]
Time system	ΔAT [s] (delta atomic time)	37.0	Ref. [27]
Satellite properties	Mass of satellite [kg] (after SDOM separation)	65.29	-
	$C_D A$ [m²]	0.80 to 7.43	-

4.2. Evaluation of SDOM De-Orbiting Capability

The results of the overall analysis are shown in Figure 8 together with the orbit altitude information of ALE-1 based on the TLE. In 2020, the orbit of ALE-1 with $C_D A = 3.0$ m² was in good agreement with the TLE. However, in 2021, the descent began to accelerate, causing the orbit to significantly change from $C_D A = 1.5$ m² to $A = 4.0$ m² over the course of the year until the SDOM was finally separated. It is worth noting that the $C_D A$ of the drag flag increased as the orbital altitude decreased. This result is a very important finding on the de-orbiting performance of a drag flag attached to a satellite in low Earth orbit (LEO). The value of $C_D A = 4.0$ m² is corresponding to a mean cross-sectional area of $A = 1.8$ m² if the drag coefficient C_D is approximated as 2.2.

As discussed in Section 3.2, the SDOM was observed to be freely moving around a single axis, fluttering like a flag. Even in this state, the drag flag delivered a de-orbiting effect on the order of $C_D A = 3.0$ m² to 4.0 m², or approximately 22 to 29% of the possible maximum value of the scenario where the drag sail is always set vertical to the velocity vector, and approximately 41 to 54% of the mean value of the scenario where the satellite is freely tumbling.

This result indicates that the shape of the de-orbiting film does not necessarily need to be fixed as a flat surface, as was regarded as the requirement so far worldwide; instead, a trailing drag flag can be about up to half as effective as a flat drag sail in LEO in the case of ALE-1 implementation. This new finding can open up breakthrough opportunities for more effective ways of implementing PMD (Post-Mission Disposal) devices in terms of their simplicity, size, mass, and reliability.

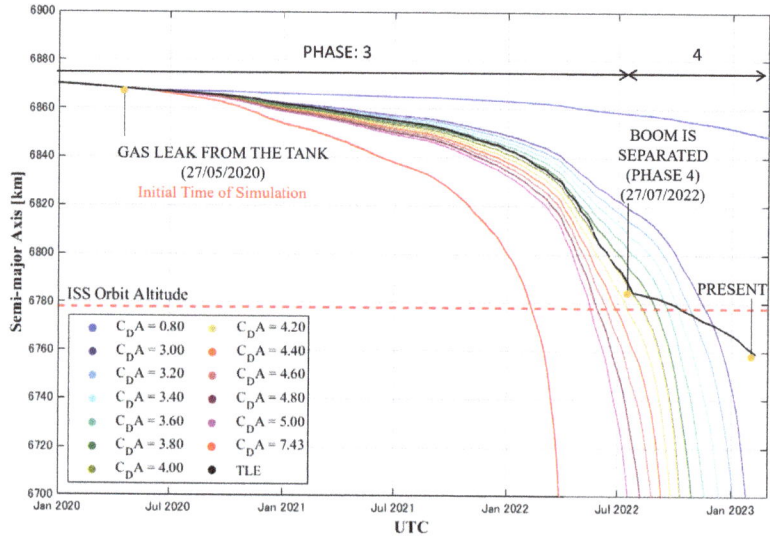

Figure 8. Comparison between the flight data and orbit propagation simulations.

The trailing drag flag is expected to move backwards with the increase in atmospheric density and is expected to stabilize in a cloak-like manner, resulting in a smaller effective cross-sectional area; in other words, ρ and A may be inversely related. The above-mentioned increasing effective cross-sectional area along the descent of the orbit can suggest that the complex behavior of the drag flag in the rarefied atmosphere may have resulted in a three-dimensional shape, which in turn resulted in the increasing mean atmospheric perturbation. It was also reported in the past study that even a flat surface that is parallel to the velocity vector delivers a non-zero atmospheric drag coefficient due to the thermochemical properties of the upper atmosphere [28]. It can be that this effect also helped the drag flag to achieve the above-mentioned de-orbiting performance. Details on this topic needs to be investigated further in the future.

5. Conclusions

ALE-1, the technology demonstration satellite for artificial shooting star generation, successfully conducted an orbital descent maneuver from an initial orbit of about 500 km down to about 400 km with the help of an SDOM system. ALE-1 also succeeded at remaining in the target orbit for a longer period to conduct its mission by separating the DOM. In addition, ALE-1 also succeeded at obtaining camera images of several different configurations of the SDOM during the orbital operation.

The SDOM, however, experienced mechanical troubles in space, and the drag sail resulted in a drag flag; nevertheless, the orbit of the ALE-1 was able to be lowered as planned, although it took longer than expected, indicating that the drag flag shape has an effective de-orbiting performance. Consequently, a comprehensive investigation was conducted on the de-orbiting capabilities of the drag flag by comparing orbital propagation simulation results and the flight data of the satellite altitude over the past three years of operation. A combination of parameters $C_D A$ was used to evaluate the de-orbiting performance of the SDOM, which can be regarded as the mean values of the drag coefficient multiplied by the cross-sectional area. In this way, the mean de-orbiting performance could be evaluated without an explicit analysis on the drag coefficient, which can be affected by the shape of the fluttering drag flag in the rarefied atmosphere in the orbit. Through the investigation, the following points can be found:

- The $C_D A$ of the drag flag increased as the orbital altitude decreased;
- The $C_D A$ of the drag flag was estimated to be ranging from 3.0 to 4.0, which corresponds to about 22 to 29% of the $C_D A$ of the fully deployed DOM and 39 to 51% of the $C_D A$ of a tumbling satellite with the deployed DOM;
- Drag flags can be effective de-orbit devices and can provide breakthroughs for future PMD devices to solve space debris problems.

The exact behaviors of the drag flag and drag sail attached to satellites are subject to further investigations; the above findings, however, indicate that materials in different shapes than sails, such as threads, tapes, and mantles, may have similar de-orbiting performances and can be utilized for future PMD devices to solve space debris problems. In addition, de-orbiting devices based on these various shapes can have possibilities to be implemented more mechanically efficiently, resulting in more lightweight and small solutions. It is therefore also possible that significantly larger and higher performance de-orbiting devices can be developed with reduced mass and envelope.

From the satellite system design point of view, the developed de-orbiting device SDOM system has several superior benefits compared with the traditional methods of de-orbiting using thruster systems. Firstly, SDOM's only 4 kg of mass is favorable for micro-satellites with a mass of approximately 100 kg or less. Secondly, it does not contain hazardous materials such as propellant and can be handled very safely on the ground, during the launch, and even possibly inside manned spacecraft. Finally, it does not necessitate active attitude control or power consumption during the de-orbiting as demonstrated in this study; hence, it can ensure that the satellite can de-orbit, even if the satellite is no longer operational after the activation of the device, which is of great benefit as a de-orbiting device. The authors sincerely hope that this research's results can contribute to the enhancement of future peaceful space utilization.

Author Contributions: Conceptualization, K.T. and T.K. (Toshinori Kuwahara); methodology, K.T.; software, K.T.; validation, T.S., Y.S., S.F., and T.K. (Toshinori Kuwahara); investigation, H.I. and T.K. (Tetsuya Kaneko); resources, Y.S. and T.K. (Tetsuya Kaneko); writing—original draft preparation, K.T.; writing—review and editing, T.K. (Toshinori Kuwahara); visualization, K.T.; supervision, T.K. (Toshinori Kuwahara); project administration, T.K. (Toshinori Kuwahara); funding acquisition, T.K. (Toshinori Kuwahara), H.I. and L.O. All authors have read and agreed to the published version of the manuscript.

Funding: This research received no external funding.

Institutional Review Board Statement: Not applicable.

Informed Consent Statement: Not applicable.

Data Availability Statement: Not applicable.

Acknowledgments: This project is supported by the Innovative Satellite Technology Demonstration Program of JAXA.

Conflicts of Interest: The authors declare no conflict of interest.

References

1. Innovative Satellite Technology Demonstration Program. Available online: http://www.kenkai.jaxa.jp/eng/research/innovative/innovative.html (accessed on 1 February 2023).
2. Morita, Y.; Imoto, T.; Tokudome, S.; Ohtsuka, H. First Launch in Months: Japan's Epsilon Launcher and Its Evolution. *Trans. JSASS Aerosp. Technol. Jpn.* **2014**, *12*, 21–28. [CrossRef] [PubMed]
3. Kuwahara, T.; Yoshida, K.; Sakamoto, Y.; Tomioka, Y.; Fukuda, K.; Fukuyama, M.; Tanabe, Y.; Shibuya, Y. Qualification results of a sail deployment mechanism for active prevention and reduction of space debries. *Proc. Int. Astronaut. Congr.* **2012**, *4*, 2565–2570.
4. Kuwahara, T.; Yoshida, K.; Sakamoto, Y.; Tomioka, Y.; Fukuda, K.; Sugimura, N. A series of de-orbit mechanism for active prevention and reduction of space debris. *Proc. Int. Astronaut. Congr.* **2013**, *3*, 2230–2234.
5. Pala, A.; Kuwahara, T.; Takeda, K.; Shibuya, Y.; Sato, Y.; Fujita, S.; Suzuki, D.; Kaneko, T. Orbital Maneuver Evaluation of Micro-satellite ALE-1 with a Separable Drag Sail. In Proceedings of the 2022 IEEE/SICE International Symposium on System Integration (SII), Narvik, Norway, 9–12 January 2022; pp. 877–881.
6. Fujita, S.; Sato, Y.; Kuwahara, T.; Sakamoto, Y.; Shibuya, Y.; Kamachi, K. Double Fail-Safe Attitude Control System for Artificial Meteor Microsatellite ALE-1. *Trans. JSASS Aerosp. Technol. Jpn.* **2021**, *19*, 9–16. [CrossRef]
7. Tangdhanakanond, P.; Kuwahara, T.; Shibuya, Y.; Honda, T.; Pala, A.; Fujita, S.; Sato, Y.; Shibuya T.; Kamachi, K. Structural Design and Verification of Aeronomy Study Satellite ALE-1. *Trans. JSASS Aerosp. Technol. Jpn.* **2021**, *19*, 42–51.
8. Konaka, M.; Fujita, S.; Sato, Y.; Shibuya, T.; Kuwahara, T.; Kamachi, K. Evaluation of thermal analysis of orbital environment of microsatellite ALE-1. In Proceedings of the 69th International Astronautical Congress: Involving Everyone, IAC 2018, Bremen, Germany, 1–5 October 2018 .
9. Shibuya, T.; Kuwahara, T.; Tangdhanakanond, P.; Shibuya, Y.; Fujita, S.; Sato, Y.; Hanyu, K.; Murata, Y.; Matsushita, T.; Kamachi, K. Thermal Design and Evaluation of the Microsatellite ALE-1. *Trans. JSASS Aerosp. Technol. Jpn.* **2021**, *19*, 821–830. [CrossRef]
10. Shibuya, Y.; Kuwahara, T.; Sato, Y.; Fujita, S.; Watanabe, H.; Mitsuhashi, Y. Orbit Design and Analysis of Artificial Meteors Generating Micro-satellites. In Proceedings of the International Astronautical Congress, IAC 2021, Dubai, United Arab Emirates, 25–29 October 2021; Volume B4.
11. Shibuya, Y.; Sato, Y.; Tomio, H.; Kuwahara, T.; Fujita, S.; Kamachi, K.; Watanabe, H. Development and Demonstration of the Mission Control System for Artificial Meteor Generating Micro-satellites. In Proceedings of the 2021 IEEE/SICE International Symposium on System Integration (SII), Iwaki, Japan, 11–14 January 2021 ; pp. 531–536.
12. Honda, T.; Kuwahara, T.; Fujita, S.; Pala, A.; Shibuya, Y.; Sato, Y.; Kamachi, K. High Precision Orbit Determination Method Based on GPS Flight Data for ALE-1. *Trans. JSASS Aerosp. Technol. Jpn.* **2021**, *19*, 744–752. [CrossRef]
13. Uto, H.; Kuwahara, T.; Honda, T. Orbit Verification Results of the De-Orbit Mechanism Demonstration CubeSat FREEDOM. *Trans. JSASS Aerosp. Technol. Jpn.* **2019**, *17*, 295–300. [CrossRef]
14. Tomioka, Y.; Yoshida, K.; Sakamoto, Y.; Kuwahara, T.; Fukuda, K.; Sugimura, N. Lessons learned on structural design of 50 kg micro-satellites based on three real-life micro-satellite projects. In Proceedings of the 2012 IEEE/SICE International Symposium on System Integration (SII), Fukuoka, Japan, 16–18 December 2012 ; pp. 319–324.
15. Kuwahara, T.; Yoshida, K.; Sakamoto, Y.; Tomioka, Y.; Fukuda, K.; Tanabe, Y.; Fukuyama, M. A sail deployment mechanism for active prevention and reduction of space debries. *Proc. Int. Astronaut. Congr.* **2011**, *3*, 2178–2184.
16. Kuwahara, T.; Yoshida, K.; Sakamoto, Y.; Takahashi, Y.; Kurihara, J.; Yamakawa, H.; Takada, A. A Japanese microsatellite bus system for international scientific missions. *Proc. Int. Astronaut. Congr.* **2011**, *5*, 3699–3706.
17. Mogi, T.; Kuwahara, T.; Uto, H. Structural Design of De-orbit Mechanism Demonstration CubeSat FREEDOM. *Trans. JSASS Aerosp. Technol. Jpn.* **2016**, *14*, 61–68. [CrossRef] [PubMed]
18. Pala, A.; Kuwahara, T.; Honda, T.; Uto, H.; Kaneko, T.; Potier, A.; Tangdhanakanond, P.; Fujita, S.; Shibuya, Y.; Sato, Y.; et al. System Design, Development and Ground Verification of a Separable De-Orbit Mechanism for the Orbital Manoeuvre of Micro-Satellite ALE-1. *Trans. JSASS Aerosp. Technol. Jpn.* **2021**, *19*, 360–367. [CrossRef]
19. Pala, A.; Kuwahara, T.; Saito, T.; Uto, H.; Shibuya, Y. Space Demonstration of Boom Extension and De-orbit Sail Deployment of the Separable De-orbit Mechanism of Micro-satellite ALE-1. *Trans. JSASS Aerosp. Technol. Jpn.* **2022**, *20*, 65–72. [CrossRef]
20. Potier, A.; Kuwahara, T.; Pala, A.; Fujita, S.; Sato, Y.; Shibuya, Y.; Tomio, H.; Tangdhanakanond, P.; Honda, T.; Shibuya, T.; et al. Time-of-Flight Monitoring Camera System of the De-orbiting Drag Sail for Microsatellite ALE-1. *Trans. JSASS Aerosp. Technol. Jpn.* **2021**, *19*, 774–783. [CrossRef]
21. Kuwahara, T.; Pala, A.; Potier, A.; Shibuya, Y.; Sato, Y.; Fujita, S.; Suzuki, D.; Kaneko, T. Orbital Demonstration of Gossamer Structure Shape Estimation using Time-of-Flight Camera System. In Proceedings of the 2022 IEEE/SICE International Symposium on System Integration (SII), Narvik, Norway, 9–12 January 2022; pp. 882–886.
22. Vallado, D.A.; McClain, W.D. *Fundamentals of Astrodynamics and Applications*, 4th ed.; Microcosm Press: Portland, OR, USA, 2013; pp. 517–729.
23. Montenbruck, O.; Gill, E. *Satellite Orbits*; Springer: Berlin/Heidelberg, Germany, 2005; pp. 53–116.
24. Tomioka, Y.; Yoshida, K.; Sakamoto, Y.; Kuwahara, T.; Fukuda, K.; Sugimura, N.; Fukuyama, M.; Shibuya, Y. Establish the environment to support cost-effective and rapid development of micro-satellites. *Proc. Int. Astronaut. Congr.* **2012**, *10*, 8470–8477.
25. Kuwahara, T.; Fukuda, K.; Sugimura, N.; Hashimoto, T.; Sakamoto, Y.; Yoshida, K. Low-Cost Simulation and Verification Environment for Micro-Satellites. *Trans. JSASS Aerosp. Technol. Jpn.* **2016**, *14*, 83–88. [CrossRef] [PubMed]

26. Picone, J.M.; Hedin, A.E.; Drob, D.P. NRLMSISE-00 empirical model of the atmosphere: Statistical comparisons and scientific issues. *J. Geophys. Res. Space Phys.* **2002**, *107*, 15–16. [CrossRef]
27. National Geospatial-Intelligence Agency Office of Geomatics. Available online: https://earth-info.nga.mil/ (accessed on 1 February 2023).
28. Fujita, K.; Noda, A. Rarefied Aerodynamics of a Super Low Altitude Test Satellite. In Proceedings of the 41st AIAA Thermophysics Conference, San Antonio, TX, USA, 22–25 June 2009; pp. 1–10.

Disclaimer/Publisher's Note: The statements, opinions and data contained in all publications are solely those of the individual author(s) and contributor(s) and not of MDPI and/or the editor(s). MDPI and/or the editor(s) disclaim responsibility for any injury to people or property resulting from any ideas, methods, instructions or products referred to in the content.

MDPI
St. Alban-Anlage 66
4052 Basel
Switzerland
www.mdpi.com

Applied Sciences Editorial Office
E-mail: applsci@mdpi.com
www.mdpi.com/journal/applsci

Disclaimer/Publisher's Note: The statements, opinions and data contained in all publications are solely those of the individual author(s) and contributor(s) and not of MDPI and/or the editor(s). MDPI and/or the editor(s) disclaim responsibility for any injury to people or property resulting from any ideas, methods, instructions or products referred to in the content.

www.ingramcontent.com/pod-product-compliance
Lightning Source LLC
LaVergne TN
LVHW070735100526
838202LV00013B/1239